Brain Source Localization Using EEG Signal Analysis

T0186377

Brain Source Localization Using EEG Signal Analysis

Munsif Ali Jatoi and Nidal Kamel

CRC Press
Taylor & Francis Group
Boca Raton London New York

CRC Press is an imprint of the
Taylor & Francis Group, an **informa** business

CRC Press
Taylor & Francis Group
6000 Broken Sound Parkway NW, Suite 300
Boca Raton, FL 33487-2742

First issued in paperback 2019

© 2018 by Taylor & Francis Group, LLC
CRC Press is an imprint of Taylor & Francis Group, an Informa business

No claim to original U.S. Government works

ISBN-13: 978-1-4987-9934-8 (hbk)
ISBN-13: 978-0-367-88497-0 (pbk)

Library of Congress Cataloging-in-Publication Data

Names: Jatoi, Munsif Ali, author. | Kamel, Nidal, author.
Title: Brain source localization using EEG signal analysis / Munsif Ali Jatoi and Nidal Kamel.
Description: Boca Raton : Taylor & Francis, 2018. | Includes bibliographical references.
Identifiers: LCCN 2017031348 | ISBN 9781498799348 (hardback : alk. paper)
Subjects: | MESH: Electroencephalography | Brain Mapping | Brain Diseases--diagnostic imaging | Brain--diagnostic imaging
Classification: LCC RC386.6.E43 | NLM WL 150 | DDC 616.8/047547--dc23
LC record available at https://lccn.loc.gov/2017031348

Visit the Taylor & Francis Web site at
http://www.taylorandfrancis.com

and the CRC Press Web site at
http://www.crcpress.com

Dedication

My grandparents: Mohammad Ali Jatoi, Sahib Khatoon Jatoi, Muhib Ali Jatoi, and Meerzadi Jatoi

Parents: Hubdar Ali Jatoi and Ghulam Fatima Jatoi

And my lovely family: Lalrukh Munsif Ali, Kazim Hussain Jatoi, and Imsaal Zehra Jatoi

With Love and Respect,
Munsif Ali Jatoi

To my beloved wife, Lama, and adorable son, Adam

Nidal Kamel

Contents

Preface

*I am not one of those whose hearts are filled with fear
when faced with the challenge to cross the deserts and
the mountains. I shall follow the pattern of the people for
whom arduous struggle is a way of life.*

LATIF (1689–1752)

The life of a researcher is full of passion to serve humanity with new ideas and target solutions that can make lives better. This is especially more evident when you are conducting research in biomedical engineering, which has a direct relation with human life betterment, and works toward identifying and solving the major issues that create hurdles in creating a healthy society. Among the various research areas in biomedical sciences, brain science has been the most attractive and developing field as many people suffer from various brain disorders globally. These disorders include epilepsy, depression, stress, schizophrenia, Alzheimer disease, and Parkinson disease. According to the World Health Organization, 1% of the world population is suffering from epilepsy, which hinders many in our society. The same is the case with other brain disorders. This field works on various aspects of the brain, which include brain modeling, brain connectivity, brain plasticity, and brain source localization. This book is written for brain science researchers, clinicians, and medical personnel with an emphasis on the field of brain source localization.

Brain source localization is a multidisciplinary field that has its roots in various fields, such as applied mathematics (bioelectromagnetism, inverse problems, Maxwell's equation, etc.); signal/image processing (basic as well as applied for various neuroimaging techniques such as magnetoencephalography/electroencephalography [MEG/EEG], functional magnetic resonance imaging, positron emission tomography); biology to understand brain anatomy; and statistics to validate the analyses from various experiments. This field emerged a few decades ago to understand human brain dynamics in a more analytical and scientific way. This advanced understanding can help society to diagnose the brain disorders mentioned above. The applications are very wide and dynamic, including research in brain source imaging, applied mathematical problems, signal/ image processing techniques, and advanced neuroimaging techniques. In addition, clinical applications include the localization of brain sources from where their origin, such as localizing epileptogenic zones for epileptic patients. Keeping in mind these constraints, this book is authored to help clinicians, researchers, and field experts in the area of brain science in general, and brain source localization in particular.

The book is divided into 10 chapters providing an introduction to the subject, neuroimaging techniques for brain analysis, detailed discussion of the EEG forward problem and the EEG inverse problem, and results obtained by applying classical (minimum norm estimation [MNE], low-resolution brain electromagnetic tomography [LORETA], beamformer) and advanced Bayesian-based multiple sparse priors (MSPs) and its modified version (M-MSP).

Chapter 1 gives insight into the field of brain source localization. Hence, the basic idea behind source localization is discussed. Furthermore, to support the introduction, sections are provided with the theory related to brain anatomy and the idea of signal generation due to any mental or physical task. Human brain anatomy is discussed to provide a basic introduction to readers to the task-oriented structure of the brain. Furthermore, the neuroimaging techniques generally used in clinics and research centers are discussed. At the end of the chapter, the economic burden due to various brain orders, and thus the potential applications of brain source analysis are covered.

Chapter 2 discusses different neuroimaging techniques in general, and provides a detailed discussion of EEG in particular. The thorough discussion on EEG includes EEG rhythms, preprocessing steps for EEG, applications of EEG, and EEG source analysis. In the source analysis section, the forward and inverse problems for brain source localization are covered. Moreover, the categorization of algorithms used for EEG-based source localization is described, which provides the foundation for the development of such algorithms. The chapter ends by listing some potential applications for EEG source localization.

Chapter 3 offers a basis for explanation of the mathematical formulation applied for the EEG forward problem. Hence, it starts with an explanation of Maxwell's equations as they are basic equations used to understand any electromagnetic phenomenon. Furthermore, the assumptions applied for brain signals are covered in the quasi-static approximation section. The dipole, which is considered as equivalent to a brain source, is defined and explained using derivations. The conductivity values for various brain regions are elaborated as provided in the literature.

Chapter 4 provides a discussion for all techniques that are usually applied for head modeling. It is observed that numerical techniques are more complex but have more resolution and good performance for source localization problems as compared with analytical methods. Thus, the finite element method (FEM), the boundary element method (BEM), and the finite difference method (FDM) are employed for head modeling to obtain a solution with high resolution for source localization. Among them, BEM is simpler as compared with FEM as it is noniterative in nature and has less computational complexity because it uses the surface as the domain rather than volume as in the case of FEM and FDM, respectively. All of these techniques are covered in this chapter along with the necessary derivations and examples.

Chapter 5 gives a detailed account of classical brain source localization techniques. A discussion is provided for a mathematical background related to the inverse problem in general, and these classical techniques in particular. Hence, MNE is defined and explained using derivations to provide a stronger base for this initial method. Furthermore, LORETA, which is an advanced version of MNE, is examined. After this, the standardized version of LORETA (i.e., sLORETA) is elaborated. The latest version of the LORETA family (i.e., exact LORETA [eLORETA]) is then covered after sLORETA. The chapter is completed by discussing the focal underdetermined system solution (FOCUSS) method, which is considered to belong to the same classical group as it employs weighted minimum norm for the source estimation.

Chapter 6 examines the hybrid techniques that were developed by mixing one of the classical techniques with another to maximize the localization capability and reduce the error. Thus initially, the hybrid weighted minimum norm (WMN) is discussed with its formulation. Moreover, WMN-LORETA is presented with its basic formulations. The discussion is continued for iterative methods based on hybridization of sLORETA and FOCUSS (i.e., recursive sLORETA-FOCUSS). Finally, shrinking LORETA-FOCUSS and its advanced version (i.e., standardized shrinking LORETA-FOCUSS [SSLOFO]) along with their major steps are explained.

Chapter 7 gives a detailed account of the subspace-based brain source localization techniques. First, subspace concepts are discussed. Linear independence and orthogonal concepts are then covered with related derivations. To explain the decomposition process for the system solution, singular value decomposition (SVD) is presented in detail. Moreover, SVD-based algorithms such as multiple signal classification (MUSIC) and recursively applied and projected-MUSIC (RAP-MUSIC) are examined in detail. Finally, the first principle vectors (FINES) algorithm is discussed to support the discussion for the subspace-based source localization algorithms.

Chapter 8 provides a detailed discussion for Bayesian framework-based inversion methods, which include MSPs and the modified version. The chapter starts with an introduction to Bayesian modeling in general. Then, Bayesian framework-based MSP is elaborated, showing that the localization efficiency is dependent on covariance matrices. The cost function (i.e., free energy) is explained along with mathematical derivations and theory. Moreover, the optimization for cost function is discussed with automatic relevance determination (ARD) and greedy search (GS) algorithms. The impact of patches on localization is explored, and thus a new method based on MSP (i.e., M-MSP) is examined. Finally, the flow is defined for the implementation of MSP.

Chapter 9 presents a thorough discussion of different aspects of the results obtained for EEG data inversion through various classical and new techniques. The results are divided into two main categories: either from synthetic data or from real-time EEG data. The synthetic data are observed for five different signal-to-noise ratio levels. A detailed discussion is provided for all methods and these methods are compared in terms of free energy, localization error (only for synthetic data), and computational time. A similar methodology is followed for real-time EEG data, where the number of individuals is kept at 10. Localization is observed for reduced electrodes with a simple mapping of 74 electrodes into seven electrodes only. However, with the reduced number of electrodes, the free energy is optimized as seen in the results. It is observed that the M-MSP is compared with classical and MSP algorithms in terms of free energy and computational complexity.

Chapter 10 summarizes the main contributions from this research work. In addition, future work is provided for researchers to gain insight into this diverse field of research. This chapter also provides directions for researchers in this area to obtain better results in the application of this knowledge to healthcare problems.

The authors are thankful to the Center for Intelligent Signal and Imaging Research (CISIR), Universiti Teknologi PETRONAS, Perak, Malaysia, for providing the necessary facilities to complete this task. The authors are also thankful to the Faculty of Engineering, Sciences and Technology, Indus University, Karachi, Sindh, Pakistan, for providing services and help for this work. We would like to extend our gratitude to our families whose patience and love have made this possible.

Finally, we welcome all comments and suggestions from readers and would love to see their feedback.

Munsif Ali Jatoi
Nidal Kamel

MATLAB® is a registered trademark of The MathWorks, Inc. For product information, please contact:

The MathWorks, Inc.
3 Apple Hill Drive
Natick, MA 01760-2098 USA
Tel: 508 647 7000
Fax: 508-647-7001
E-mail: info@mathworks.com
Web: www.mathworks.com

Authors

Munsif Ali Jatoi, PhD, earned a PhD in electrical and electronic engineering from the Universiti Teknologi PETRONAS, Perak, Malaysia, in 2016. Prior to this, he earned an MSc in advanced photonics and communications from the University of Warwick, United Kingdom, and a BE in electronics) from Mehran University of Engineering and Technology, Jamshoro, Pakistan, in 2009 and 2007, respectively. Dr. Jatoi has more than eight years of teaching experience locally and internationally as an associate professor, assistant professor, lecturer, and graduate assistant, respectively. He has 25 research publications in journals and conferences to his credit. He has presented his research work in various international exhibitions and won two silver medals for his performance in ITEX (International Invention, Innovation & Technology Exhibition) and SEDEX (Science and Engineering Design Exhibition) in Malaysia. Dr. Jatoi has filed five patents in the field of EEG (electroencephalography) source localization and has coauthored a book chapter with Taylor & Francis. His research interests are brain signal processing, EEG inverse problem, epilepsy prediction, brain connectivity, and applied mathematics for neuroscience. Currently, Dr. Jatoi is serving as an associate professor at the Faculty of Engineering, Science and Technology (FEST), Indus University, Karachi, Sindh, Pakistan.

Nidal Kamel, earned a PhD (Hons) from the Technical University of Gdansk, Poland, in 1993. Since 1993, he has been involved in research projects related to estimation theory, noise reduction, optimal filtering, and pattern recognition. He developed a single-trial subspace-based technique for ERP (event-related potential) extraction from brain background noise, a time-constraints optimization technique for speckle noise reduction in SAR (synthetic-aperture radar) images, and introduced a data glove for online signature verification. His current research interest is mainly in EEG (electroencephalography) signal processing for localization of brain sources, assessment of cognitive and visual distraction, neurofeedback, learning and memory recall, in addition to fMRI–EEG (functional magnetic resonance imaging–electroencephalography) data fusion. He is the editor of *EEG/ERP Analysis: Methods and Applications*, CRC Press, New York, 2015. Currently, he is an associate professor at the PETRONAS University of Technology, Perak, Malaysia. He is an IEEE (the Institute of Electrical and Electronics Engineers) senior member.

List of symbols

$\mathbf{A}\dagger$	Moore–Penrose pseudo-inverse of $\mathbf{A}$
A^{-1}	Inverse of A
$A^\wedge$	Estimate of A
arg	Argument
det (A)	Determinant of A
c	Speed of light (3×10^8)
Hz	Hertz, cycles per second
Exp(.)	Exponential
$\mathbf{I}$	Identity matrix
Im(.)	Imaginary part
j	$\sqrt{-1}$
KL	Kullback–Leibler
max	Maximum
min	Minimum
$p(\mathbf{Y})$	Probability of $\mathbf{Y}$
$p(\mathbf{J}\|\mathbf{Y})$	Probability of event $\mathbf{J}$ given event $\mathbf{Y}$
p_x	Probability density function of x
$\Re^M$	M-dimensional space
Re(.)	Real part
subcorr	Subspace correlation
$\|x\|_2$	Euclidean or L-2 norm of x
$\|x\|_F$	Frobenius norm of x
$\langle \cdot, \cdot \rangle$	Inner product
$\dfrac{dy}{dx}$	Differentiation of y with respect x
$\dfrac{\partial y}{\partial x}$	Partial differentiation of y with respect to x
$\displaystyle\int f(x)dx$	Integration of function $f(x)$ with respect to x
$\mathbf{E}[.]$	Statistical expectation
Var (.)	Variance operator
*	Linear convolution
∇	Del or Nabla operator

$(.)^T$	Transpose operator
$(.)^H$	Hermitian; complex conjugate transpose
∞	Infinity
$\propto$	Sign of proportionality
$\sum_{i=1}^{N}$	Summation of N components
$\forall n$	For all n values
$\exists x$	There exists an x
ω	Angular frequency in radians per second
α	Penalty term
α	Alpha brain rhythm
β	Beta brain rhythm
γ	Gamma brain rhythm
δ	Delta brain rhythm
θ	Theta brain rhythm
$\geq$	Greater than or equal to
$\leq$	Smaller than or equal to
$\approx$	Approximately equal to

List of abbreviations

3D	Three-Dimensional
AAS	Average Artefact Subtraction
ADD	Attention Deficiency Disorder
ADHD	Attention Deficit Hyperactivity Disorder
AED	Antiepileptic Drugs
Ag–AgCl	Silver–Silver Chloride
AP	Action Potential
ARD	Automatic Relevance Determination
BC	Boundary Conditions
BCG	Ballistocardiogram
BCI	Brain–Computer Interface
BEM	Boundary Element Method
BOLD	Blood Oxygenation Level Dependent
BSS	Blind Source Separation
CCA	Canonical Correlation Analysis
CNS	Central Nervous System
CSF	Cerebrospinal Fluid
CT	Computed Tomography
DTI	Diffusion Tensor Imaging
ECD	Equivalent Current Dipole
EEG	Electroencephalography
ECG	Electrocardiography
eLORETA	Exact Low-Resolution Brain Electromagnetic Tomography
EMG	Electromyography
EM	Expectation Maximization
EMF	Electromagnetic Field
EMD	Empirical Mode Decomposition
EOG	Electrooculography
EPSP	Excitatory Postsynaptic Potential
ERP	Event-Related Potential
ESPRIT	Estimation of Signal Parameters via Rotational Invariance Techniques

FDM	Finite Difference Method
FEM	Finite Element Method
FFT	Fast Fourier Transform
FINES	First Principle Vectors
fMRI	Functional Magnetic Resonance Imaging
FOCUSS	Focal Underdetermined System Solution
FVM	Finite Volume Method
GA	Genetic Algorithm
GLM	Generalized Linear Model
GS	Greedy Search
HEOG	Horizontal Electrooculography
HWMN	Hybrid Weighted Minimum Norm
ICA	Independent Component Analysis
LORETA	Low-Resolution Brain Electromagnetic Tomography
LORETA-FOCUSS	Low-Resolution Brain Electromagnetic Tomography–Focal Underdetermined System Solution
MDD	Major Depression Disorder
MEG	Magnetoencephalography
MM	Millimeter
MMF	Magnetomotive Force
MNE	Minimum Norm Estimation
MRC	Medical Research Council
MS	Millisecond
MSP	Multiple Sparse Priors
MUSIC	Multiple Signal Classification
OBS	Optimal Basis Set
PET	Positron Emission Tomography
PCA	Principle Component Analysis
PSD	Power Spectral Density
PSP	Postsynaptic Potentials
qEEG	Quantitative Electroencephalography
RAP-MUSIC	Recursively Applied and Projected-Multiple Signal Classification
ROOT MUSIC	Root Multiple Signal Classification
ReML	Restricted Maximum Likelihood
sMRI	Structural Magnetic Resonance Imaging
SNR	Signal-to-Noise Ratio
SSLOFO	Standardized Shrinking Low-Resolution Brain Electromagnetic Tomography Focal Underdetermined System Solution
STM	Short-Term Memory
SVD	Singular Value Decomposition

SVM	Support Vector Machine
VEOG	Vertical Electrooculography
WHO	World Health Organization
WT	Wavelet Transform
USD	U.S. Dollar

chapter one

Introduction

The field of brain source localization using electroencephalography (EEG)/ magnetoencephalography (MEG) signals emerged a few decades ago in neuroscience for a variety of clinical and research applications. To solve the EEG inverse problem, the forward problem needs to be solved first. The forward problem suggests the modeling of the head using advanced mathematical formulations [1]. Thus, head modeling for the solution of the forward problem is categorized as either analytical or numerical [2]. The numerical head modeling schemes have proved more efficient in terms of resolution provided, which leads to a better solution for the inverse problem. The commonly used techniques are the boundary element method (BEM), finite element method (FEM), and finite difference method (FDM). Among these methods, BEM is simpler and noniterative as compared with FEM and FDM. However, FEM has higher computational complexity as well as better resolution by covering more regions and allowing efficient computation of irregular grids [3,4].

Different neuroimaging techniques are used to localize active brain sources. However, when EEG is used to solve this problem, it is known as the EEG inverse problem. The EEG inverse problem is an ill-posed optimization problem as unknown (sources) outnumbers the known (sensors). Hence, to solve the EEG ill-posed problem, many techniques have been proposed to localize the active brain sources properly—that is, with better resolution [5]. The inverse techniques are generally categorized as either parametric approaches or imaging approaches [6]. The parametric methods assume the equivalent current dipole representation for brain sources. However, the imaging methods consider the sources as intracellular currents within the cortical pyramidal neurons. Hence, a current dipole is used to represent each of many tens of thousands of tessellation elements on the cortical surface. Thus, the source estimation in this case is linear in nature, as the only unknowns are the amplitudes of dipoles in the tessellation element. However, because the number of known quantities (i.e., electrodes) is significantly less than the number of unknowns (sources that are >10 K), the problem is underdetermined in nature. Hence, regularization techniques are used to control the degree of smoothing.

Mathematically, the EEG inverse problem is typically an optimization problem. Thus, there are a variety of minimization procedures, which include Levenberg–Marquardt and Nelder–Mead downhill simplex searches, global optimization algorithms, and simulated annealing [7]. However, all equivalent current dipole models apply principle component analysis (PCA) and singular value decomposition (SVD) to obtain a first evaluation related to the number and relative strength of field patterns existing in data before the application of a particular source model. According to the categorization discussed above, a number of techniques based on least squares principle, subspace factorization, Bayesian approaches, and constrained Laplacian have been developed for brain source localization. The methods are generally based on least squares estimation with minimum norm estimates (MNEs) [8–10] and its modified form with Laplacian smoothness, such as low-resolution brain electromagnetic tomography (LORETA), standardized LORETA (sLORETA), and exact LORETA (eLORETA) [11–14]; beamforming approaches [15]; and some parametric array signal processing-based subspace methods, such as multiple signal classification (MUSIC), recursive MUSIC, recursively applied and projected-MUSIC (RAP-MUSIC), and estimation of signal parameters via rotational invariance technique (ESPRIT) [16–19]. The Bayesian framework-based methods are known as multiple sparse priors (MSP), which is the latest development in the field of brain source localization. This technique is discussed in detail in the literature [20–23].

These methods are characterized according to certain parameters such as resolution, computational complexity, localization error, and validation. Some of these methods (LORETA, sLORETA, eLORETA) have low spatial resolution. However, array signal processing-based MUSIC and RAP-MUSIC offer better resolution but at the cost of high computational complexity. In addition, some other methods such as a focal underdetermined system solution (FOCUSS) provides a better solution with high resolution; however, due to heavy iterations in the weight matrix, FOCUSS has high computational complexity [24]. Some hybrid algorithms are also proposed for this purpose, such as weighted minimum norm–LORETA (WMN-LORETA) [25], hybrid weighted minimum norm [26], recursive sLORETA-FOCUSS [27], shrinking LORETA-FOCUSS [28], and standardized shrinking LORETA-FOCUSS (SSLOFO) [29]. These hybrid algorithms provide better source localization but have a large number of iterations, thus resulting in heavy computational complexity. Besides, the techniques that are hybridized with LORETA and sLORETA suffer from low spatial resolution. Hence, the limitations of algorithms developed with any techniques include less stable solution, more computational burden, blurred solution, and slow processing.

After presenting a brief background on the field of brain source localization, we shall move on to unearth more about this dynamic and multidimensional research field. This research field—that is, brain source localization—involves brain signal/image processing, optimization algorithms, mathematical manipulations, and applied computational techniques. And a large variety of applications exist for this field, such as in brain disorder (epilepsy, brain tumor, schizophrenia, depression, etc.) applications, behavioral science applications, psychological studies, and traumatic brain injury applications. These topics are covered in detail later.

1.1 Background

This section provides a detailed introduction for the human brain anatomy, modern neuroimaging techniques, which are used to capture image/signal data from the brain through different means, and overall economic burden due to brain disorders for which EEG source localization is used.

1.1.1 Human brain anatomy and neurophysiology

The human brain is the most complex organ with 10^{12} neurons, which are interconnected via axons and dendrites, and 10^{15} synaptic connections. This complex structure allows it to release/absorb quintillion of neurotransmitter and neuromodulator molecules per second. The metabolism of the brain can be analyzed through radioactively labeled organic molecules or probes that are involved in processes of interest such as glucose metabolism or dopamine synthesis [30]. Brain development starts at a primary age of 17–18 weeks of parental development and generates the electrical signals until death [31]. Neurons act as processing units for the brain activity to send or receive signals from/to various parts of the body to the brain. According to embryonic developments, the human brain can be divided into three regions anatomically: the forebrain (or *prosencephalon*), the midbrain (or *mesencephalon*), and the hindbrain (or *rhombencephalon*) [32]. However, broadly, the brain tissues are categorized as either gray matter or white matter. The brain surface is divided into four lobes: the frontal lobe, the parietal lobe, the temporal lobe, and the occipital lobe. A detailed discussion about brain anatomy and related functionality is provided in the following sections.

The brain is the most important and complex part of the central nervous system. Composed of various neurons that act as the data-processing unit for the brain, it only weighs around 1500 g [33]. Brain tissues are categorized as either gray matter or white matter. The brain regions formed by gray matter are responsible for processing information and establishing connections with white matter. The gray matter is mostly composed of unmyelinated neurons. The white matter is composed of myelinated neurons, which are used as connectors to the gray matter. Because myelinated neurons transmit nerve signals faster, white matter functions to increase the speed of signal transmission between the connections.

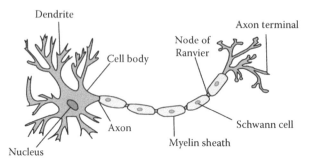

Figure 1.1 Basic neuron structure.

The brain cells are either neurons or neuroglia. Neurons act as a processing unit for the brain, sending/receiving information from/to various parts of the body to the brain. Although the sizes of neurons in the brain vary, their basic functional unit remains the same. Each neuron contains a cell body, which is called a soma; a nucleus; a dendrite tree; and an axon [34]. The dendrites, which originate from the cell body (or soma) and branch repeatedly, are used for the reception of inputs from other neurons. The synapse is formed by the branching of an axon end. The synapse is an information highway between two neurons. The basic structure of a neuron is shown in Figure 1.1 [35].

Electrical signals are received by each neuron and are processed accordingly. However, neuroglia or glial cells are helping agent for neurons. They just support and protect neurons. There are four types of glial cells: astrocytes, oligodendrocytes, microglia, and ependymal cells. Figure 1.2 shows a neuron in culture with synapses visualizing the activity [36].

According to embryonic developments, the human brain can be divided into three regions anatomically: the forebrain (or *prosencephalon*), the midbrain (or *mesencephalon*), and the hindbrain (or *rhombencephalon*). The forebrain consists of the cerebrum, thalamus, hypothalamus, and pineal gland. The cerebral area is usually called the *telencephalon*, and the whole area of the thalamus, hypothalamus, and pineal gland is called the *diencephalon* [37]. The diencephalon is located in the midline of the brain; however, the telencephalon or the cerebrum is the most superior structure, which has the lateral ventricles, basal ganglia, and cerebral cortex. By contrast, the midbrain or mesencephalon is located at the center of the brain exactly between the pons and the diencephalon. It is further divided into the tectum and cerebral peduncles. The forebrain or prosencephalon is composed of telencephalon and diencephalon, and conversely the *diencephalon* (or interbrain) includes the thalamus, hypothalamus, and pineal glands. The thalamus consists of a pair of oval

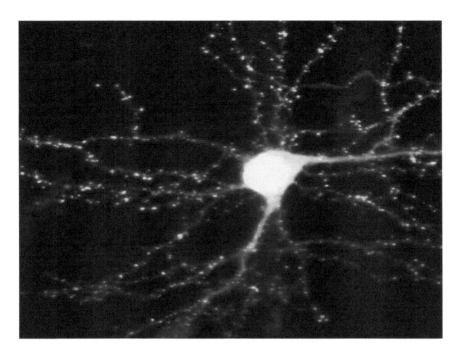

Figure 1.2 A living neuron in culture: Green dots indicate excitatory synapses and red dots indicate inhibitory synapses. (From G. G. Gross et al., *Neuron*, vol. 78(6), pp. 971–985, 2003. With permission.)

masses of gray matter lower to the lateral ventricles and surrounding the third ventricle. The thalamus plays a vital role in learning by sending sensory information for processing and in memory processing. The hypothalamus is located lower to the thalamus and acts as the controller of the brain for body temperature, hunger, thirst, blood pressure, heart rate, and production of hormones. The pineal glands are located posterior to the thalamus in a small region called the epithalamus. They produce the hormone melatonin. The amount of melatonin produced is related to age. As a person ages, the amount of melatonin produced decreases and hence sleep is reduced, as sleep is directly related to the hormone (melatonin) produced. The main and largest part of the forebrain is the cerebrum, which controls the main functions of the brain such as language, logic, reasoning, and creative activities. The location of the cerebrum is around the diencephalon and superior to the cerebellum and brainstem. Figure 1.3 shows the brain structure [38].

The cerebrum is divided into two hemispheres (Figure 1.4), namely, the left hemisphere and the right hemisphere. The line dividing the cerebrum is known as the longitudinal fissure, which runs midsagittal down the center of cerebrum. Both hemispheres are divided into four lobes:

Telencephalon   Cerebellum

Diencephalon   Pons

Mesencephalon   Medulla

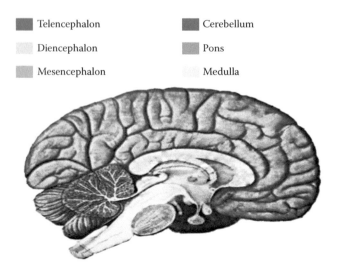

Figure 1.3 Main anatomical features of the brain.

frontal, parietal, temporal, and occipital. The left hemisphere is dedicated to rational tasks such as language, computation, and analysis, whereas the right hemisphere is for visual and spatial perception and intuition. The surface of the cerebrum is known as the cerebral cortex, which is responsible for most of the processing tasks in the cerebrum. This cortex is composed of gray matter only [39].

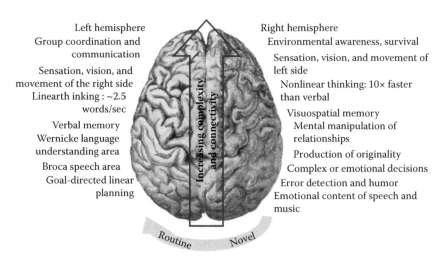

Left hemisphere
Group coordination and communication
Sensation, vision, and movement of the right side
Linearth inking : ~2.5 words/sec
Verbal memory
Wernicke language understanding area
Broca speech area
Goal-directed linear planning

Right hemisphere
Environmental awareness, survival
Sensation, vision, and movement of left side
Nonlinear thinking: 10× faster than verbal
Visuospatial memory
Mental manipulation of relationships
Production of originality
Complex or emotional decisions
Error detection and humor
Emotional content of speech and music

Increasing complexity and connectivity

Routine   Novel

Figure 1.4 Brain hemispheres showing the corresponding functions. (From F. Morrissette. Available: http://lobe.ca/en/non-classee/two-ears-two-cerebral-hemispheres/#.VptCMVKtH8m. Accessed on October 23, 2017. With permission.)

A summary of functionality of the four lobes (frontal, parietal, temporal, and occipital) of the left and right hemispheres is provided below.

Frontal lobe

- Behavior, emotions, parts of speech, reasoning, creative thinking
- Judgment, reasoning, problem solving
- Speaking and writing
- Movement
- Intellectual thinking, attentiveness, self-awareness

Parietal lobe

- Language interpretation, words
- Sensory intelligence (sense of touch, pain, temperature, etc.)
- Interpretation of signals from visual, audio stimuli
- Spatial and visual perception

Temporal lobe

- Language perception
- Memory
- Listening
- Sequencing and organization

Occipital lobe

- Visual processing (light, color, movement)

These lobes are separated by fissures, which are present in all lobes. As a result, the distinction between various lobes can be judged by visual inspection. Figure 1.5 shows the structural analysis for various lobes with fissure labeling. It can be visualized that the frontal and parietal lobes are separated through the central fissure, and the temporal and parietal lobes are separated by the sylvian fissure, and so on.

After presenting a brief overview of the brain anatomy, the neurophysiology of the brain is discussed to understand how the electrical signals are generated to produce electromagnetic activity inside the brain, which is supposed to be localized.

The electrical signals that are measured by neuroimaging techniques are produced as a result of bioelectromagnetism inside the brain. This electromagnetic field is produced by ion currents inside the brain. As described in the previous sections, neurons are brain cells that transmit/receive information. However, transmitting and receiving information is

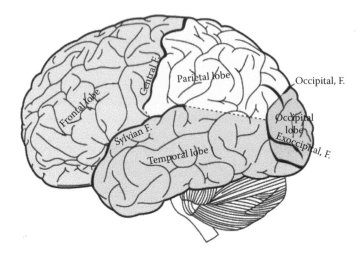

Figure 1.5 Lobes of the cerebral cortex. (From H. Gray, *Anatomy of the Human Body*, Philadelphia, PA: Lea & Febiger, 1918; Bartleby, 2000. With permission.)

dependent on the rise/fall of electrical potentials at the cell membrane. Hence, this potential difference is responsible for the generation of currents flowing into and outside the neuron [40]. Therefore, the signals in the dendrites are termed as postsynaptic potentials (PSPs) and the signals emitted, moving along the axon, are said to be action potentials (APs). APs are the information transmitted by a nerve. APs are generated due to the exchange of ions across the neuron membrane and are a temporary change in membrane potential, which is transmitted along the axon. The lifetime of AP is 5–10 ms with amplitude of 70–100 mV. Figure 1.6 shows an illustration for APs [31].

In the resting state, a neuron has a negative potential of −60 mV compared with extracellular environment. This potential is dependent on the synaptic activity. If an AP is traveling along the line, and ending in an excitatory synapse, an excitatory PSP is generated in the following neuron. However, with the increase in the potential (which is due to multiple traveling potentials), a certain threshold for the membrane potential is reached. If the fiber ends in an inhibitory synapse, then hyperpolarization results, indicating the presence of inhibitory PSP. Because of the production of inhibitory PSP, a flow of cations is generated, which causes a variance in cell membrane potential [41,42]. The current is produced due to this flow through the extracellular space and is recorded through EEG recording devices. The frequencies of such signals are very low (in the range of 100 Hz). The time interval of PSP is large (10–20 ms) with lower amplitude (0.1–10 mV) as compared with APs.

Because less amplitude of fields is generated by APs and PSPs, it is necessary to calculate their summation to measure them directly through

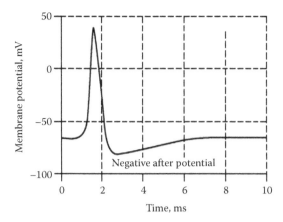

Figure 1.6 Action potential. (From S. Sanei and J. A. Chambers, *EEG Signal Processing*. Hoboken, NJ: John Wiley & Sons, 2013. With permission.)

EEG. However, APs offer low temporal duration, and thus it is difficult to summate them. Therefore, PSPs are better to generate measurable electromagnetic fields outside the head to be measured by EEG electrodes. Hence, it is observed that approximately 10^{14} neurons should be generated simultaneously to produce a voltage measurable by electrodes [43]. In addition, it is notable that only certain types of neurons can generate a sufficient amount of potential that can be recorded via electrodes. Such cells include pyramidal neurons, which are located in the gray matter of the cortex which have thick dendrites perpendicular to the cortex; by contrast, stellate cells possess dendrites in all directions and are unable to produce a significant potential that can be measured via electrodes on the scalp. Hence, the EEG signal that is recorded using electrodes is nothing but a measurement of the currents that flow when synaptic excitation of the dendrites occurs in pyramidal neurons within the cerebral cortex.

1.1.2 Modern neuroimaging techniques for brain disorders

Studies in neuroscience are performed to understand the activation of neurons in the brain, which leads to cognitive processes. This process of understanding brain activation is a very complex and multidisciplinary phenomenon because it involves the combination of neuroscience (intense study of brain anatomy revealing connections and dynamic interactions between synaptic micro sources) and deep-applied mathematical skills for brain signaling and an imaging approach specialized for neuroimaging. As a result, several methods have been developed to analyze brain activity by taking advantage of the combined features provided by the described disciplines. Therefore, the techniques that are used for the study of brain activity with various features are termed *functional neuroimaging techniques*. These neuroimaging techniques analyze electromagnetic signals generated in the brain, which are responsible for brain activation during various activities. Based on various parameters, such as ease of use, availability, resolution, and computational complexity, the most popular neuroimaging approaches used for clinical and research purposes are MEG, EEG, positron emission tomography, functional magnetic resonance imaging (fMRI), and near-infrared spectroscopy. Each method has specific properties related to temporal and spatial resolution, and thus has different modalities and clinical applications.

According to the literature, noninvasive imaging techniques, such as MRI/positron emission tomography/fMRI, have good spatial resolution and are helpful to understand the brain dynamics for different kinds of activations. However, these techniques have less temporal resolution when compared with EEG/MEG. In addition, the aforementioned techniques are expensive in terms of maintenance and availability. Hence, EEG/MEG techniques are used at the clinical and research levels to record electric and magnetic activities of the brain to better understand cognitive and behavioral functions.

EEG was introduced as a neuroimaging technique by German psychiatrist, Hans Berger, in 1924 [44]. It measures the brain's electrical activity using sensors placed on the scalp. The EEG/MEG recordings are taken by using a set of electrodes (19, 128, 256, etc.) by following the standard procedure as provided. Hence, the process to localize active sources (electrical generators), which are responsible for the measured electric/magnetic fields, is termed as *brain source localization*. If the neuroimaging technique adopted to take the measurements is EEG, then this is termed as *EEG source localization* or *EEG inverse problem*.

This book discusses EEG as a neuroimaging technique only. Hence, the detailed discussion for EEG, its analysis in general and source analysis in particular, its preprocessing, and other related topics are provided in next chapters.

1.1.3 Economic burden due to brain disorders

The effort to understand the brain source localization problem began approximately 40 years ago by correlating the existing brain's electrophysiological information with the basic physical principles controlling the volume currents in conductive media. The information acquired through the brain source localization based on EEG signals is helpful to diagnose different brain disorders such as epilepsy, schizophrenia, depression, Alzheimer disease, and Parkinson disease.

Among them, the most common is epilepsy; according to World Health Organization (WHO) statistics, around 50 million people in the world have been diagnosed as epileptic [45]. This figure amounts to 40%–70% per 0.1 million people in developing countries, which is quite an alarming situation. Furthermore, more than 80% of cases are reported in developing countries, with three-fourths of them not receiving proper treatment. Epilepsy accounts for 0.5% of the global burden on economy and growth. In some countries, such as India, the cost per case has been reported as US$344. However, the social effects on the life of an epileptic individual vary from one country to another. This includes the restriction on marrying, and on going to restaurants, theaters, clubs, and other public buildings. Epilepsy can be treated by antiepileptic drugs or by surgical therapy. Hence, brain source localization techniques are used to localize epileptogenic regions where the abnormal behavior of neurons can be observed during a seizure. In addition, the localization information can help clinicians to operate on brain tumors, as this is one of the reasons behind secondary (or *symptomatic*) epilepsy.

Another important brain disorder is schizophrenia, which affects more than 21 million people around the globe. It is characterized by disruptions in thinking, affecting language, perception, and losing the sense of self. This disorder develops in late adolescence or in early adulthood. The major drawbacks of this chronic disorder include drowsiness, dizziness with changing positions, blurred vision, rapid heartbeat, skin rashes, and so forth [46]. This psychiatric disorder is studied through pilot programs in developing countries, and it is observed that proper healthcare facilities and medication can help reduce it.

In a previous study [47], brain source localization methods such as LORETA and sLORETA were applied to determine the region of abnormality for schizophrenic patients. In the analysis, EEG data were used, and the corresponding power maps were obtained using the LORETA software package. These power maps were compared to show differences between schizophrenic patients and controls. In another study [48], a low-resolution source localization technique, eLORETA, was used to define functional connectivity and source localization for schizophrenic patients. Furthermore, blind source separation techniques are used for localization of P300 sources in schizophrenia patients [49].

A hybrid algorithm combining features of both EEG and fMRI for source localization has also been seen in the literature for various brain disorders including schizophrenia, dementia, and depression.

Depression is a common brain disorder characterized by sadness, loss of interest or pleasure, feelings of guilt, disturbed sleep, appetite loss, and so on. It can be categorized as mild or severe. In the severe case, suicidal attempts are reported. It is reported that brain source localization techniques (such as LORETA) are applied to localize the position of the region most affected by major depression disorder. In Coutin-Churchman and Moreno [50], LORETA was used to localize the powers associated with various bands (alpha, theta, and beta) for patients with and without depression. Hence, in this way, it can be deduced that brain source localization can be used for the recognition and subsequent treatment of major depression disorder.

1.1.4 Potential applications of brain source localization

As discussed in the previous section, brain source localization has a variety of applications in the field of brain science. Its main application is for various brain disorders such as epilepsy, schizophrenia, depression, stress, Parkinson disease, and tumor analysis. Thus, according to statistics and healthcare data, it is evident that source localization is a helpful tool to analyze and diagnose brain disorders when presented in hospitals and brain research centers. The most potential applications for source localizations are as follows:

- The system developed to localize the active brain sources can be used by hospitals to help surgeons and physicians operate on patients with various brain disorders. This system includes forward modeling through numerical methods such as FEM, FDM, and BEM, and inverse methods (MNE, LORETA/sLORETA/eLORETA, or modern Bayesian-based multiple sparse prior), which have features of low localization error and less computational complexity.
- The developed system can be used by researchers in the field of neuroscience in particular and signal/image processing experts in general. The researchers can be from a variety of fields, such as from applied mathematics, optimization experts, signal/image processing experts, and brain science experts.
- The product commercialization can be initiated from the developed system, which can be used in hospitals for proper diagnoses for patients with brain disorders. Thus, the commercialized product can give optimum benefit to researchers and medical personnel.

Besides these advantages, source analysis can be applied to behavioral studies such as psychological analysis, sleep disorders to localize the location of the affected brain part, or measurement of driver distraction quantitatively.

Summary

This chapter provided a brief insight into the field of brain source localization. First, a brief background was provided to discuss the general idea behind source estimation. The subsequent sections described theories related to brain anatomy, and explained signal generation during any mental or physical task. The anatomical structure of the human brain was then briefly discussed to provide insight to the reader into the task-oriented structure of the brain. Furthermore, neuroimaging techniques that are frequently used in hospitals were discussed. Finally, the economic burden resulting from various brain orders and the potential applications of brain source analysis were covered. Thus, this chapter provided an in-depth introduction into source localization.

References

1. M. S. Hämäläinen and J. Sarvas, Realistic conductivity geometry model of the human head for interpretation of neuromagnetic data, *IEEE Transactions on Biomedical Engineering*, vol. 36, pp. 165–171, 1989.
2. H. Hallez et al., Review on solving the forward problem in EEG source analysis, *Journal of Neuroengineering and Rehabilitation*, vol. 4, p. 1, 2007.
3. J. J. Ermer, J. C. Mosher, S. Baillet, and R. M. Leahy, Rapidly recomputable EEG forward models for realizable head shapes, *Journal of Physics Medicine and Biology*, vol. 46, pp. 1265–1281, 2001.
4. H. Buchner, G. Knoll, M. Fuchs, A. Rienacker, R. Beckmann, M. Wagner, J. Silny, and J. Pesch, Inverse localization of electric dipole current sources in finite element models of the human head, *Electroencephalography and Clinical Neurophysiology*, vol. 102, pp. 267–278, 1997.
5. G. H. Golub and V. Pereyra, The differentiation of pseudo-inverses and nonlinear least squares problems whose variables separate, *SIAM Journal on Numerical Analysis*, vol. 10, pp. 413–432, 1973.
6. S. Baillet, J. C. Mosher, and R. M. Leahy, Electromagnetic brain mapping, *IEEE Signal Processing Magazine*, vol. 18, pp. 14–30, 2001.
7. M. A. Jatoi, N. Kamel, A. S. Malik, I. Faye, J. M. Bornot, and T. Begum, EEG-based brain source localization using visual stimuli, *International Journal of Imaging Systems and Technology*, vol. 26, pp. 55–64, 2016.
8. M. S. Hämäläinen and R. J. Ilmoniemi, Interpreting magnetic fields of the brain: Minimum norm estimates, *Medical and Biological Engineering and Computing*, vol. 32, pp. 35–42, 1994.
9. K. Uutela, M. S. Hämäläinen, and E. Somersalo, Visualization of magnetoencephalographic data using minimum current estimates, *NeuroImage*, vol. 10, pp. 173–180, 1999.
10. M. Hämäläinen, R. Hari, R. J. Ilmoniemi, J. Knuutila, and O. V. Lounasmaa, Magnetoencephalography—Theory, instrumentation, and applications to noninvasive studies of the working human brain, *Reviews of Modern Physics*, vol. 65, p. 413, 1993.
11. R. D. Pascual-Marqui, C. M. Michel, and D. Lehmann, Low resolution electromagnetic tomography: A new method for localizing electrical activity in the brain, *International Journal of Psychophysiology*, vol. 18, pp. 49–65, 1994.
12. R. D. Pascual-Marqui, Review of methods for solving the EEG inverse problem, *International Journal of Bioelectromagnetism*, vol. 1, pp. 75–86, 1999.
13. R. Pascual-Marqui, Standardized low-resolution brain electromagnetic tomography (sLORETA): Technical details, *Methods and Findings in Experimental Clinical Pharmacology*, vol. 24, pp. 5–12, 2002.
14. R. D. Pascual-Marqui, Discrete, 3D distributed linear imaging methods of electric neuronal activity. Part 1: Exact, zero error localization, 2007.

15. Y. Jonmohamadi, G. Poudel, C. Innes, D. Weiss, R. Krueger, and R. Jones, Comparison of beamformers for EEG source signal reconstruction, *Biomedical Signal Processing and Control*, vol. 14, pp. 175–188, 2014.

16. J. C. Mosher and R. M. Leahy, Recursive MUSIC: A framework for EEG and MEG source localization, *IEEE Transactions on Biomedical Engineering*, vol. 45, pp. 1342–1354, 1998.

17. J. C. Mosher and R. M. Leahy, Source localization using recursively applied and projected (RAP) MUSIC, *IEEE Transactions on Signal Processing*, vol. 47, pp. 332–340, 1999.

18. T. K. R. Roy, ESPRIT estimation of signal parameters via rotational invariance techniques, *IEEE Transactions on Acoustics, Speech and Signal Processing*, vol. 37, pp. 984–995, 1989.

19. J. C. Mosher, R. M. Leahy, and P. S. Lewis, EEG and MEG: Forward solutions for inverse methods, *IEEE Transactions on Biomedical Engineering*, vol. 46, pp. 245–259, 1999.

20. K. Friston et al., Multiple sparse priors for the M/EEG inverse problem, *NeuroImage*, vol. 39, pp. 1104–1120, 2008.

21. J. López, V. Litvak, J. Espinosa, K. Friston, and G. R. Barnes, Algorithmic procedures for Bayesian MEG/EEG source reconstruction in SPM, *NeuroImage*, vol. 84, pp. 476–487, 2014.

22. J. D. Lopez, G. R. Barnes, and J. J. Espinosa, Single MEG/EEG source reconstruction with multiple sparse priors and variable patches, *Dyna*, vol. 79, pp. 136–144, 2012.

23. M. A. Jatoi, N. Kamel, J. D. López, I. Faye, and A. S. Malik, MSP based source localization using EEG signals. In *Intelligent and Advanced Systems (ICIAS), 2016 6th International Conference on*, pp. 1–5. IEEE, 2016.

24. I. F. Gorodnitsky, J. S. George, and B. D. Rao, Neuromagnetic source imaging with FOCUSS: A recursive weighted minimum norm algorithm, *Electroencephalography and Clinical Neurophysiology*, vol. 95, pp. 231–251, 1995.

25. R. Khemakhem, W. Zouch, A. Taleb-Ahmed, and A. Ben Hamida, A new combining approach to localizing the EEG activity in the brain: WMN and LORETA solution, in *International Conference on BioMedical Engineering and Informatics, 2008 (BMEI 2008)*, New York: IEEE, 2008, pp. 821–824.

26. C. Y. Song, Q. Wu, and T. G. Zhuang, Hybrid weighted minimum norm method: A new method based LORETA to solve EEG inverse problem, in *27th Annual International Conference of the Engineering in Medicine and Biology Society, 2005 (IEEE-EMBS 2005)*, New York: IEEE, 2005, pp. 1079–1082.

27. K. Rafik, B. H. Ahmed, F. Imed, and T.-A. Abdelmalik, Recursive sLORETA-FOCUSS algorithm for EEG dipoles localization, in *First Workshops on Image Processing Theory, Tools and Applications, 2008 (IPTA 2008)*, New York: IEEE, 2008, pp. 1–5.

28. L. He Sheng, Y. Fusheng, G. Xiaorong, and G. Shangkai, Shrinking LORETA-FOCUSS: A recursive approach to estimating high spatial resolution electrical activity in the brain, in *First International IEEE EMBS Conference on Neural Engineering, 2003*, New York: IEEE, 2003, pp. 545–548.

29. L. Hesheng, P. H. Schimpf, D. Guoya, G. Xiaorong, Y. Fusheng, and G. Shangkai, Standardized shrinking LORETA-FOCUSS (SSLOFO): A new algorithm for spatio-temporal EEG source reconstruction, *IEEE Transactions on Biomedical Engineering*, vol. 52, pp. 1681–1691, 2005.

30. S. R. Cherry and M. E. Phelps, *Imaging Brain Function with Positron Emission Tomography, Brain Mapping: The Methods*, New York, NY: Academic Press, 1996, pp. 191–221.

31. S. Sanei and J. A. Chambers, *EEG Signal Processing*, Hoboken, NJ: John Wiley & Sons, 2013.

32. InnerBody. Available: http://www.innerbody.com/image/nerv02.html#full-description. Accessed on January 16, 2017.

33. R. Plonsey, *Bioelectric Phenomena*, Hoboken, NJ: Wiley Online Library, 1969.

34. D. A. Drachman, Do we have brain to spare? *Neurology*, vol. 64, pp. 2004–2005, 2005.

35. M. Flagg. Available: http://hubpages.com/education/Structure-of-a-Neuron. Accessed on January 16, 2017.

36. G. G. Garrett et al., Recombinant probes for visualizing endogenous synaptic proteins in living neurons, *Neuron*, vol. 78, pp. 971–985, 2013.

37. Mayfield Brain and Spine. Available: http://www.mayfieldclinic.com/PE-AnatBrain.htm#.U6uV7rH6Xxg. Accessed on January 16, 2017.

38. F. Morrissette. Available: http://lobe.ca/en/non-classee/two-ears-two-cerebral-hemispheres/#.VptCMVKtH8m. Accessed on January 16, 2017.

39. H. Gray, *Anatomy of the Human Body*, Philadelphia, PA: Lea & Febiger, 1918; Bartleby, 2000.

40. C. Koch, *Biophysics of Computation: Information Processing in Single Neurons*, Oxford, UK: Oxford University Press, 1998.

41. E. J. Speckmann, Introduction of the neurophysiological basis of the EEG and DC potentials, *Electroencephalography: Basic Principles, Clinical Applications, and Related Fields*, 5th edn., E. Niedermeyer and F. Lopes da Silva (eds.), Philadelphia, PA, US: Lippincott Williams & Wilkins, pp. 15–26, 1993.

42. G. M. Shepherd, *The Synaptic Organization of the Brain*, London, UK: Oxford University Press, 1974.

43. S. Murakami and Y. Okada, Contributions of principal neocortical neurons to magnetoencephalography and electroencephalography signals, *Journal of Physiology*, vol. 575, pp. 925–936, 2006.

44. A. Massimo, In Memoriam Pierre Gloor (1923–2003): An appreciation, *Epilepsia*, vol. 45, p. 882, 2004.

45. World Health Organization. Available: http://www.who.int/en/. Accessed on January 16, 2017.

46. National Institute of Mental Health. Available: http://www.nimh.nih.gov/health/topics/schizophrenia/index.shtml. Accessed on January 16, 2017.

47. T. Miyanishi, T. Sumiyoshi, Y. Higuchi, T. Seo, and M. Suzuki, LORETA current source density for duration mismatch negativity and neuropsychological assessment in early schizophrenia, *PLoS One*, vol. 8, p. e61152, 2013.

48. L. Canuet et al., Resting-state EEG source localization and functional connectivity in schizophrenia-like psychosis of epilepsy, *PLoS One*, vol. 6, p. e27863, 2011.

49. L. Spyrou, M. Jing, S. Sanei, and A. Sumich, Separation and localisation of P300 sources and their subcomponents using constrained blind source separation, *EURASIP Journal on Applied Signal Processing*, vol. 2007, pp. 89–89, 2007.

50. P. Coutin-Churchman and R. Moreno, Intracranial current density (LORETA) differences in QEEG frequency bands between depressed and non-depressed alcoholic patients, *Clinical Neurophysiology*, vol. 119, pp. 948–958, 2008.

chapter two

Neuroimaging techniques for brain analysis

Introduction

This chapter deals with the discussion concerning neuroimaging techniques that are used to understand the brain's physiological and cognitive dynamics. Thus, the chapter first provides an introduction to overall functional brain imaging and its related applications. The chapter then moves on to describe briefly famous neuroimaging techniques such as functional magnetic resonance imaging (fMRI), magnetoencephalography (MEG), and positron emission tomography (PET). However, because we are here to discuss EEG-based source localization, EEG as a neuroimaging technique is discussed in detail with its particular application in brain source localization. After this, both forward and inverse problems for EEG source localization are described. Hence, the potential applications of EEG source localization are provided to learn about solving this ill-posed problem. Finally, a list of optimization algorithms that are used to localize the active brain sources is given.

2.1 fMRI, EEG, MEG for brain applications

Over the past 40 years, many neuroimaging techniques have been developed. Currently, these techniques have advanced significantly and are used for various healthcare applications, such as functional brain imaging, which begins with improvement in understanding the basic mechanism of cognitive processes. This will lead to better characterization of pathologies, which can be utilized to cure brain disorders such as epilepsy, schizophrenia, depression, Alzheimer disease, and stress. This field is multidisciplinary and involves applied mathematics, optimization, signaling, imaging, and photonics. Thus, it is possible to have a better understanding of the human brain through noninvasive imaging techniques involving electrophysiological, hemodynamic, metabolic, and neurochemical processes, which occur in both the normal and pathological human brain. Each of these imaging methods has specific properties related to temporal and spatial resolution. However, they are utilized for clinical purposes according to their usage and modality. The neural processes in a normal human brain are understood through these powerful imaging modalities. However, clinically, they are applied to understand and treat serious neurological and neurophysiological disorders.

The analysis of brain metabolism is carried out through radioactively labeled organic molecules or probes that are involved in the process of glucose metabolism or dopamine synthesis. The imaging for dynamic change in spatial distribution of such probes when they are transported and altered chemically within the brain is performed through PET. The spatial resolution for PET imaging is 2 mm; however, the temporal resolution depends on process dynamics, which is under study, and the noise produced by photon counting. Thus, the temporal resolution can vary for up to several minutes. Apart from this modality, local hemodynamic changes can be recorded for neural activity assessment in a more precise and useful way. PET, fMRI, and transcranial optical imaging can be used to record the hemodynamic changes that occur due to neuron activation and induction of localized changes in blood flow and oxygenation levels. Among them, fMRI is the most powerful functional neuroimaging modality as the MRI scanner used for scanning can provide both anatomical and functional features with high spatial resolution, but at a high instrumentation cost. fMRI comes with different clinical MRI magnets (1.5 , 3 , 5 , and 7 T), with higher magnetic fields resulting in higher resolution and improved signal-to-noise ratio (SNR) values. fMRI has a higher spatial resolution (range, 1–3 mm), but its temporal resolution is low (1 s), which is due to slow hemodynamic response as compared with neural activity. Besides the low temporal resolution, fMRI has the disadvantage of difficult interpretation of its data due to the complex relationship between blood oxygenation level-dependent (BOLD) changes, which are detected by fMRI imaging and ongoing neural activity. Hence, the direct comparison between BOLD activation levels and neural activity cannot be performed.

MEG measures the magnetic induction outside the head due to brain electrical activity [1]. The temporal resolution offered by MEG is higher than fMRI and PET because it measures the induction directly, unlike fMRI, which depends on BOLD activation levels [2]. However, the spatial resolution of MEG is limited by the small number of spatial measurements—that is, few hundreds in the case of MEG, which for PET or fMRI is in the range of tens of thousands [3]. In the case of MEG and EEG, the brain is treated as a current generator. The MEG measurement is obtained using a sensor head array system that monitors the brain's magnetic activity. The main concern with MEG recordings is the noise, which is also the case with EEG, which is discussed later. The noise caused due to instruments used for MEG recordings can be reduced using superconductive materials [4]. The noise contributed by radiofrequency waves can be removed using a shielded environment made up of successive layers

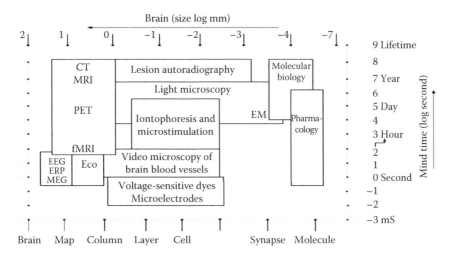

Figure 2.1 Overview of neuroimaging techniques with respect to temporal and spatial resolution. CT, computed tomography; ERP, event-related potential; fMRI, functional magnetic resonance imaging; MEG, magnetoencephalography; MRI, magnetic resonance imaging; PET, positron emission tomography.

of copper, aluminum, and so on. Besides high-frequency noises, the low-frequency noises are reduced by using a gradiometer for sensing the signals. MEG has the same potential applications as EEG—that is, to be used for clinical and research purposes for analyzing healthy and impaired brains. However, one of the distinct advantages of using MEG is that it is sensitive to primary currents, unlike EEG, which is sensitive to secondary or volume current. One of the major applications of MEG is for brain source localization, which has the same mathematical formulation as for EEG. In addition, it should be noted that the basic formulation for head modeling using MEG to solve the forward problem related to brain source localization is also similar to that using EEG. Hence, detailed discussions pertaining to inversions used for EEG are also applicable for MEG. The basic difference between EEG and MEG is of measuring quantity; that is, the former measures electrical activity of the brain by placing a set of electrodes on the scalp, whereas the latter measures magnetic field with a sensor cap. Figure 2.1 shows the comparison between various neuroimaging techniques in terms of different parameters.

Since the inception of such modern neuroimaging techniques, remarkable changes have been noted in acquisition modules/methods, data preprocessing, data handling, and signal processing techniques. This development has started a new era not only for clinicians but also for researchers, engineers, physicists, and mathematicians to develop and discover underlying principles for brain signal analysis through M/EEG and fMRI. Such advances in signal-processing methods, which include fast Fourier transform (FFT), frequency-domain analysis of EEG time series, power spectral density (PSD), and phase coherence, have provided bases for the foundation of understanding brain dynamics for cognitive and mental tasks [5]. This foundation along with the advancements in anatomical MRI data for providing access to individual brain anatomy laid the foundation for functional localization of M/EEG activity in 90 s. This field of study for understanding the brain's physiological, mental, and functional abnormalities using M/EEG is termed as M/EEG source localization. This problem has to deal with estimation of the location and distribution of the current sources responsible for the electromagnetic activity inside the brain. After a brief discussion related to various neuroimaging techniques, we now move to EEG, which is the major technique covered in this book.

2.1.1 EEG: An introduction

EEG is defined as "The noninvasive/invasive neuroimaging technique having high temporal, low spatial resolution which records the brain activity by measuring electrical signals generated by pyramidal neurons in cortical region of brain with the help of variety of electrode arrangement." EEG is a neuroimaging technique that was developed by German physicist and psychiatrist Hans Berger in 1924, to measure brain activity with a set of electrodes. The set of electrodes is placed on the surface of the scalp to measure the electrical potential of the patient under observation [6]. EEG can also be defined as a functional neuroimaging technique with high temporal resolution (in milliseconds), which measures potential differences as linear functions of source strengths and nonlinear functions of dipole locations [7]. The EEG recordings can be used for direct, real-time monitoring of spontaneous and evoked brain activity, which allows for spatiotemporal localization of neuronal activity.

In the early years of development, EEG signals were recorded using galvanometers. However, with the advancement in modern electronic circuitry, signals are now recorded using a set of electrodes, differential amplifiers for each channel, and filters. With the increase in the number of channels, the computational complexity is increased proportionally. Hence, the latest emerging techniques are utilized to analyze the EEG data recorded through a set of electrodes. The overall quality of the EEG signals is dependent on many factors among which the type of electrode is important. According to the literature, different types of electrodes are used for EEG recording, such as disposable (gel-less and pregelled) electrodes, reusable disc electrodes, headbands and electrode caps, saline-based electrodes, and needle electrodes [8]. The electrodes that are normally used for clinical and research applications are composed of Ag–AgCl material with a diameter less than 3 mm. These electrodes

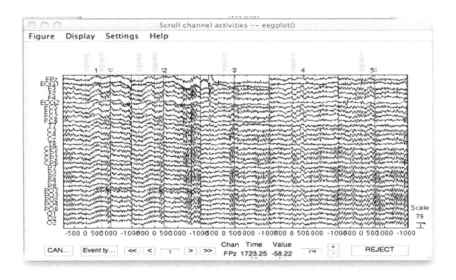

Figure 2.2 Sample EEG recording signals.

are attached to the amplifier using long wires so that small magnitude signals acquired from the scalp can be amplified. For invasive applications, needle electrodes are planted inside the skull with minor operations [9]. To better record EEG signals, the electrode impedance is maintained between 1 and 5 KΩ. Sample EEG signals are shown in Figure 2.2 [10].

In 1958, a standard electrode placement system for the measurement of brain signals using EEG was proposed [11–13]. This standardization was based on putting landmarks on head anatomy. The system so developed was called the 10–20-electrode placement system. This system was approved by the International Federation of Societies for Electroencephalography and Clinical Neurophysiology initially for only 21 electrodes. In this setting, odd electrodes are placed on the left and even on the right. For an increased number of electrodes (128, 256, 512, etc.), one can simply place the additional electrode with equidistance between the electrodes.

A diagrammatic view for the 10–20-electrode system showing distance measurements for various head regions with 75 electrodes along with reference electrodes is shown in Figure 2.3. It is common practice that the electrodes connected to ear lobes (A1 and A2) are taken as reference electrodes. However, in modern instrumentation, the choice of a reference does not play a significant role in the measurement [14]. For such systems, other references such as FPz, hand, or leg electrodes may be used [15]. The standard nomenclature is as follows: A = earlobe, O = occipital, P = parietal, F = frontal, and Fp = frontal polar.

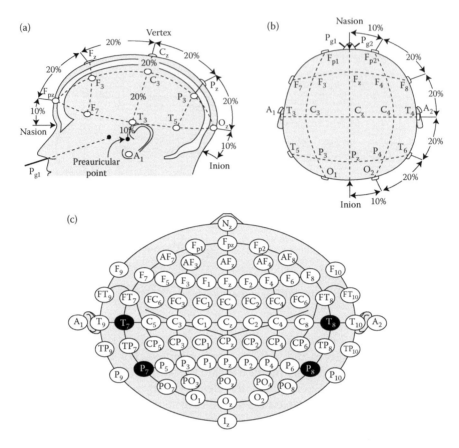

Figure 2.3 Diagrammatic view of the 10–20-electrode system. (a) and (b) show three-dimensional measures, and (c) shows a two-dimensional view of the electrode setup [5]. (From S. Sanei and J. A. Chambers. *EEG Signal Processing*. John Wiley & Sons, 2013. With permission.)

Apart from the regular number of electrodes used for signal acquisition, sometimes extra electrodes are needed to measure electrooculography (EOG), electrocardiography, and electromyography for muscle recordings. For specified applications, a single-channel EEG is taken. These examples include cognitive task detection and monitoring, finger tapping (brain–computer interface), event-related potentials (ERPs), and so on. There are also two arrangements for electrodes: bipolar or unipolar. In the bipolar arrangement, electrical potential is measured between a pair of electrodes. However, in the unipolar arrangement, the potential of each electrode is either compared with a reference/neutral electrode or the average of all electrodes is taken.

2.1.1.1 EEG rhythms

EEG signals are continuously changing waveforms that show the potential difference in oscillating fashions. The overall trend of EEG signals with respect to their amplitudes and patterns is dependent on the excitation of the brain. The amplitudes measured on the scalp are in the range of microvolts with frequencies ranging from 0.5 to 100 Hz. The visual inspection of EEG patterns is helpful for the diagnoses of various brain disorders and states. For healthy individuals, the values of amplitude and frequency vary in a set pattern. For this, the brain waves are classified according to the frequency band in which they reside. There are five important groups formed on the bases of variant frequency range, namely, alpha (α), theta (θ), beta (β), delta (δ), and gamma (γ), respectively. According to Sanei and Chambers [5], Berger introduced the alpha and beta waves in 1924. However, the gamma waves were introduced by Jasper and Andrews [16].

The delta wave was reported by Walter in 1936 [17], and theta (θ) waves were reported in Sterman et al. [18]. The following section provides a brief introduction for each rhythm.

Alpha waves

1. Have a frequency range of 8–13 Hz with an approximate sinusoidal structure.
2. Have an amplitude less than 50 μV.
3. Are found in normal persons in a wake state having less exciting work.
4. Are present mostly in the occipital region.
5. Are present during the eyes-closed resting position during ERP experiments.

Beta waves

1. Have a frequency range of 14–26 Hz.
2. Have an amplitude less than 30 μV.
3. Are found in normal persons with active thinking, concentration, and in an excited state.
4. Are present mainly in the frontal and central regions.
5. A high-level beta wave is generated when a healthy individual is in a panic situation.

Gamma waves

1. Have a frequency range over 30 Hz.
2. Have a very low amplitude due to infrequent occurrence.
3. Are found during ERP tasks for special cases and diseases.
4. Are located in the frontocentral region of the brain.

Theta waves

1. Have a frequency range of 4–7 Hz.
2. Have a very low amplitude, which is less than 100 μV.
3. Are found during sleep or in emotional stress conditions.
4. Are present in the parietal and temporal regions.
5. Are used for stress study analysis.

Delta waves

1. Have a frequency range of 0.5–4 Hz.
2. Have an amplitude less than 100 μV.
3. Are found during deep sleep or in serious brain disease.
4. Are present in the central cerebrum mostly in the parietal lobe.
5. Are mostly confused with muscle artifact signals.

The EEG signals associated with these rhythms are shown in Figure 2.4.

Besides these basic brain rhythms, some researchers have quoted some minor rhythms as well, which are a phi rhythms (having range less than 4 Hz) occurring at the eyes-closed position [19]. The kappa rhythm

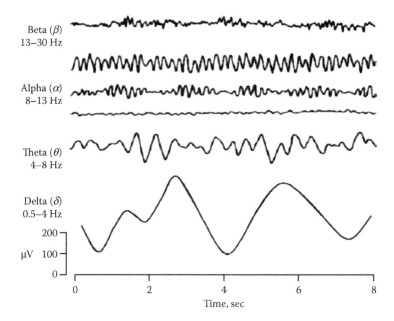

Figure 2.4 Four brain waves observed in EEG signal analysis. (From S. Sanei and J. A. Chambers. *EEG Signal Processing*. John Wiley & Sons, 2013. With permission.)

is an alpha-like rhythm that occurs at the anterior temporal lobe. This rhythm is produced due to eye movement and is considered an artifact. The others have sigma activity lying in the frequency range of 11–15 Hz, tau rhythm, chi rhythm, and lambda rhythm, and so on [20–22]. Although rhythm identification with naked eye inspection is possible, it sometimes gives errors, resulting in a failure to identify the disease correctly. Therefore, advanced digital signal processing tools for EEG signal analysis have been developed to separate the bands correctly.

2.1.1.2 Signal preprocessing

The EEG signal acquired from the human scalp with the help of electrodes using the 10–20-electrode placement system contains different types of noises (pulse artifact, muscles artifact, line noise, eye movement/blink artifact, etc.), which cause hindrances to using the EEG data for specific applications [23]. Major noise that corrupts the EEG data is from eye blink, which is a natural phenomenon. Other noises include line noise, electronic thermal noise, ballistocardiogram (BCG) artifact, and so on. In addition, during EEG–fMRI simultaneous recordings, the EEG signals are severely disturbed due to variation in magnetic and electric fields because of change in oxygen levels and corresponding sensitivity of the oxygen molecule to the magnetic field [3]. This artifact is known as BCG artifact. Therefore, the raw EEG signal contaminated by these noises and artifacts is subjected to preprocessing to make it useful for further analysis of brain disorders and cognitive problem issues. Such noises are removed by estimation through linear and nonlinear adaptive filtering methods [24,25]. However, there are certain artifact removal methods for EOG and BCG artifacts, which are explained briefly in the following sections.

2.1.1.2.1 Filtering and artifact removal The EEG signals that are measured through electrodes are in the range of <30 Hz frequency. All the band (alpha, beta, gamma, theta, delta) activity lies within this region. However, any frequency component above this range (such as an electromyography artifact due to muscle activity when the frequency is >30 Hz) is removed using low-pass filters. In either case, where it is needed to remove low-frequency components (such as baseline shift signals having frequency <0.1 Hz), a high-pass filter is employed. In some cases where the data-acquisition system is not able to cancel out the effect of 50-Hz line noise, a notch filter is used to mitigate this effect. The equalizer filters are used to compensate for the nonlinearities in the recording system in the frequency response of amplifiers used. The sample signals for artifacts are shown in Figure 2.5.

Many techniques are used to cancel out the effect of eye blink and BCG artifacts. For the removal of such noises, an adaptive noise cancellation

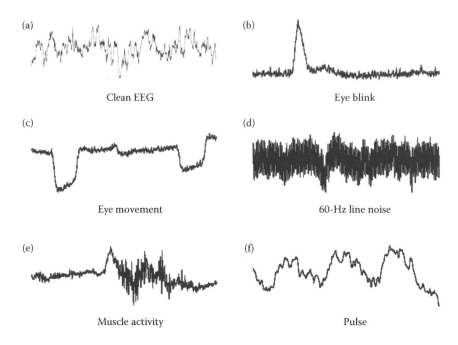

(a) Clean EEG

(b) Eye blink

(c) Eye movement

(d) 60-Hz line noise

(e) Muscle activity

(f) Pulse

Figure 2.5 Artifact sample signals. (From S. Sanei and J. A. Chambers. *EEG Signal Processing.* John Wiley & Sons, 2013. With permission.)

technique is often used. In this technique, a reference signal is created with approximately the same statistical features as the noise or artifact signal. For EOG artifact removal, some researchers have proposed a regression-based analysis (for time and frequency domains) to estimate propagation factors in [25,26]. Principal component analysis and support vector machine are also used by researchers, but these methods are dependent on the uncorrelatedness of EEG signals and EOGs. The other approaches used to remove EOG artifacts are independent component analysis, canonical correlation analysis, empirical mode decomposition, blind source separation, wavelet transform, and so on.

However, for the removal of BCG artifact, several algorithms have been developed so far. These techniques are broadly categorized to fall either in the time or in the frequency domain. In the temporal domain approach, the BCG artifact template is extracted channel wise, whereas in the spatial domain case, the BCG artifact is estimated statistically. The early techniques recorded for BCG removal were average artifact subtraction algorithms described in Sun and Hinrichs [26]. Later, various methods, such as principal component analysis, empirical mode decomposition, and optimal basis sets were developed to remove this artifact effectively from the EEG data for further analysis [27].

2.1.1.2.2 Segmentation The EEG signal is considered a nonstationary signal as it exhibits various statistical parameters for each segment. For example, in alert conditions, a normal individual will be recorded with strong alpha wave activity unlike in eye-blink or other states with different statistical parameters (such as Gaussianity check, skewness, kurtosis, negentropy) being recorded. Therefore, an EEG signal is divided into small segments having similar statistical (temporal and spatial) characteristics to provide an ease for diagnosis and analysis. Hence, the segmented signals are considered as stationary to some extent. The length of each segment is usually from 1 second to several minutes depending on the data and analysis to be carried out. For the varying lengths of each segment, a nonlinear technique (such as adaptive segmentation) is used. A very common example of EEG data segmentation is epileptic data segmentation where the individuals' data are divided into preictal, ictal, and postictal segments, respectively.

2.1.1.3 Applications of EEG

EEG has diversified applications in the neuroscience community, which can be categorized as for research, clinical, and brain computer interfacing (BCI) purposes. For research purposes, EEG is normally used in biomedical centers where modern signal and image processing techniques along with advanced mathematical approaches (adaptive linear and nonlinear filtering, subspace techniques, blind source separation methods, etc.) are applied to a variety of brain-related problems. These problems include source localization, stress monitoring, depression disorder study, drug addiction monitoring, memory recall process, quantitative analysis of EEG (which is termed as qEEG) for tumor applications and brain injury, multimodal data fusion (EEG/fMRI), neurofeedback, EEG artifact removal, and so on.

For clinical purposes, EEG can be used for the detection and analysis of many brain disorders, including both psychiatric and cognitive disorders. The most important of all is epilepsy, as 1% of the world population is suffering from this disorder. The other disorders include dementia, schizophrenia, depression, stress analysis, Alzheimer disease, Parkinson disease, attention deficit hyperactivity disorder and attention deficit disorder, delirium, and so on. EEG analysis is also used to analyze drug addiction and its corresponding effects on the individual's brain rhythms. Memory retention and recalling analysis is also an application of EEG for short-term memory analysis among patients with lose memory records. The other most common disorders that are analyzed using patients' EEG data are amnestic disorder (or amnesia), substance-related disorder, mood disorder, anxiety disorder, somatoform disorder, dissociative disorder, sexual disorder, gender identity disorder, eating disorders, sleep disorders, dyslexia, impulse-controlled disorder, and personality disorder [28].

Another potential application of EEG is for BCI-related cognitive applications. BCI is a communication system between brain signals and a computer, which is operated by signals generated by the human brain. The signals can be EEG or ERP signals depending on the particular application, for example, the drivers' alert system, fatigue analysis, and system design (such as wheelchair control using BCI) for handicap persons [29]. The factors that are studied for BCI systems include boredom, fatigue, stress, and other related illness. Some of the studies report the usage of EEG for sport sciences as well [30].

2.1.2 EEG source analysis

The estimation of the location and distribution of the current sources responsible for the electromagnetic activity inside the brain based on the potential recorded through the electrodes is one of the major problems in EEG. This problem is known as source localization or EEG inverse problem as the data (potentials) are given and one has to design the model from the available data. In other words, given a set of electric potentials from discrete sites on the surface of the head and the associated positions of those measurements and the geometry and conductivity of regions within the head, the location and magnitude of the current sources within the brain are calculated [31]. The purpose of this source localization is to reconstruct the current distribution in the human brain using potential differences or magnetic fluxes measured with the help of electrodes from the scalp. This source modeling by EEG for noninvasive localization of epileptogenic zones helps in clinical applications such as for surgery in patients with partial seizures [32]. The localization information for the active sources in the brain helps to diagnose pathological, physiological, mental, and functional abnormalities related to the brain. Therefore, EEG source localization has been an active area of research for decades. In the past few years, the source localization method, due to its use in clinical applications such as that for epileptic surgery, has produced more than 150 research publications. These publications are based on a software-based mathematical solution for the EEG inverse problem. However, less than half of the publications addressed the clinical validation for investigation of focal epilepsy [33]. The efforts to understand the localization problem began 40 years ago by correlating the existing body of electrophysiologic knowledge about the brain with the basic physical principles controlling the volume currents in conductive media [34–37].

The fundamental idea of source localization is divided into two main problems: one is termed a forward problem and the other is an inverse problem. The forward problem is the prediction of potential differences from the current sources inside the brain, and the estimation of the location of the sources from the measured data is termed as inverse problem. Hence, the forward problem is dedicated to measuring the data parameters from the available model, whereas the inverse searches for the best approximation to design the model from the available data parameters. The following sections explain in detail the forward problem and the inverse problem in general, especially for EEG.

2.1.2.1 Forward and inverse problems

In the physical world, if the data values are extracted/estimated from the given model by applying some physical theories to the model, then the problem is said to be a modelization problem, simulation problem, or forward problem [38]. This is a straightforward procedure that requires fewer computations, with less errors because the model with complete description is with us. However, the inverse problem suggests predicting the model using the available measured parameters.

In the physical world, a finite amount of data is available to reconstruct a model with infinitely many degrees of freedom. Hence, the inverse problem is not unique, and there are many models that can explain the data equally well. By contrast, the forward problem has a unique solution. As an example taken from [39], consider measurements of the gravity field around a planet: given the distribution of mass inside the planet, we can uniquely predict the values of the gravity field around the planet (forward problem). However, there are different distributions of mass that give exactly the same gravity field in the space outside the planet. Therefore, the inverse problem of inferring the mass distribution from observations of the gravity field has multiple solutions (in fact, an infinite number). Because of this, in the inverse problem, one needs to explicitly consider *a priori* information on the model parameters. One also needs to be careful in the representation of the data uncertainties.

The inverse problem is nonunique in nature; that is, many models can fit the data. In addition, the estimated data are tainted with errors, so the estimated model always differs from the true model. The model with finite degrees of freedom is termed a discrete model, and one with continuous data and infinite degrees of freedom is termed a continuous model. The model's estimation and model's appraisal are different for both systems. According to Hadamard [39], if a physical problem has a solution with uniqueness and stability in it, then the inverse problem is assumed to be well posed; otherwise, it is termed as ill-posed. A majority of geophysical problems are ill-posed, which means that they have nonuniqueness and instability. Therefore, it is assumed that the predicted model is just an approximation of the true model. The inversion problem consists of two steps, namely, an estimation problem and an appraisal problem [39]. Let the true data be denoted by d, true model denoted by m, and the estimated model by $m\sim$, and data being tainted by error. In this case, one can assume the inverse problem as a combination of what is to estimate and a relationship between estimated and true, which is said to be appraisal.

Therefore,

$$\text{Inverse Problem} = \text{Estimated Problem} + \text{Appraisal Problem}$$

There are several methodologies used to address the inverse problem. Some of them are listed below.

2.1.2.1.1 Methodologies in inverse problems The methods used to solve an inverse problem are as follows:

- *Deterministic inverse problems*: In such methods, regularization is needed as they have a worst-case convergence with infinite dimensional space. However, there is no assumption of noise to be made on measurements.
- *Statistics*: In these types of ill-posed problems, estimators are applied for the solution. The dimension is finite with noise assumed to be randomly distributed for the measurements.
- *Bayesian inverse problems*: For these inverse problems, *posteriori* distribution is maximized through any optimization algorithm to obtain unknown parameters. The dimension is assumed to be finite. However, the prior and noise are assumed with any distribution with most of the cases as Gaussian.
- *Control theory*: These inverse problems deal with infinite dimensional space. Thus, there is no assumption taken for noise for the system measurements.

Following the discussion of an ill-posed problem in general, now we move on to discuss the EEG inverse problem in particular.

2.1.2.1.2 Inverse problems for EEG The EEG source localization is an underdetermined ill-posed inverse problem because the number of unknown parameters is greater than the number of known parameters. In general, there are two approaches to solving the inverse problem for EEG signals: parametric and imaging methods. For the parametric method, it is assumed that sources are represented by a few equivalent current dipoles of unknown locations and moments. These dipoles are then supposed to be estimated through the nonlinear numerical method. By contrast, the imaging methods are based on the assumption that primary sources are intracellular currents in the dendritic trunks of the cortical pyramidal neurons, which are aligned normally to the cortical surface. Hence, a current dipole is assigned to each of many tens of thousands of tessellation elements on the cortical surface with the dipole orientation constrained to

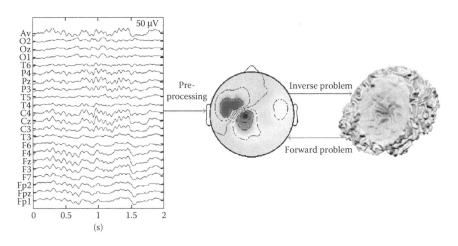

Figure 2.6 Forward and inverse problems for EEG data. (From K. Wendel, O. Väisänen, J. Malmivuo et al. EEG/MEG source imaging: Methods, challenges, and open issues, *Computational Intelligence and Neuroscience*, vol. 2009, Article ID 656092, p. 12, 2009. doi:10.1155/2009/656092. With permission.)

equal the local surface normal. In this case, the inverse problem becomes linear, as the only unknown quantity is dipole amplitude.

The least square source estimation by applying various minimization techniques is categorized as a parametric method. The minimization methods include Levenberg–Marquardt and Nelder–Mead downhill simplex searches to global optimization schemes using multistart methods, genetic algorithms, and simulated annealing.

There are many methods for obtaining the solution to the EEG inverse problem, which can be categorized according to the methodology adopted for implementation. Some of the methods are defined alone and some are defined independently. However, other methods include hybrid methods, which are formed by mixing existing algorithms with other algorithms for yielding better results with less errors and accurate localization of sources within the brain. The following section provides a detailed discussion for various inverse solutions in the field of EEG source localization.

Figure 2.6 gives a pictorial view of forward and inverse problems [6].

2.1.3 Inverse solutions for EEG source localization

The inverse solution for EEG source localization depends on various parameters such as resolution, localization error, computational time, and overall system complexity. Thus, various methods are applied to solve this ill-posed problem. Some of them are based on least square optimization

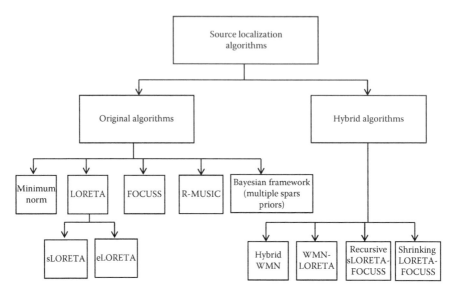

Figure 2.7 Flow diagram of inversion techniques utilized for source localization.

(minimum norm estimation), others are based on second-order Laplacian (low-resolution brain electromagnetic tomography [LORETA], standardized LORETA, and exact LORETA), and some are based on singular value decomposition, such as multiple signal classification (MUSIC) or recursively applied and projected MUSIC (RAP-MUSIC). However, the latest one is multiple sparse priors, which is based on the Bayesian framework. Some of the methods are also reported in literature that are hybrid in nature, such as weighted minimum norm–LORETA (WMN-LORETA) and recursive standardized LORETA–focal underdetermined system solution (LORETA-FOCUSS). Based on this discussion, it is clear that there are several methods to solve the inverse problem related to EEG. Thus, this section provides a detailed discussion related to all methods along with their mathematical derivations and implementation for data. Figure 2.7 shows the classification for different inversion techniques.

A detailed explanation along with necessary mathematical derivations and physical interpretations for both forward and inverse problems will be provided in the upcoming chapter.

2.1.4 Potential applications of EEG source localization

Brain source localization using EEG signals has significant applications for various medical and clinical applications. Some such potential applications are as follows:

- The localization algorithm can be used by hospitals to help surgeons and physicians in operating on patients with brain disorders, such as localizing epileptogenic zones for patients having epilepsy, tumor localization, and localizing the affected areas for other brain disorders.
- The localization algorithm can be used by researchers in the field of neuroscience in particular and signal processing experts in general for better advancement in the healthcare systems.
- A product can be commercialized from the algorithm, which can be launched in the neuromarket so that hospitals, researchers, and experts can get optimum benefit from it.

Summary

This chapter briefly discussed the various neuroimaging techniques and provided an in-depth discussion for EEG. The detailed discussion for EEG includes the EEG rhythms, the preprocessing steps for EEG, applications of EEG, and then EEG source analysis. In the "EEG Source Analysis" section, we discussed forward and inverse problems in general and then for brain source localization in particular, respectively. Furthermore, the categorization of algorithms used for EEG-based source localization is explained, which defines the foundation for development of such algorithms. Thus, the methods are listed in a flow diagram according to their categorization. These methods are covered in detail with their mathematical background and physical interpretation later in this text. Finally, some of the potential applications for EEG source localization are listed.

References

1. F. Lopes da Silva, Functional localization of brain sources using EEG and/ or MEG data: Volume conductor and source models, *Journal of Magnetic Resonance Imaging*, vol. 22(10), pp. 1533–1538, 2004.
2. B. W. Muller, M. Juptner, W. Jentzen, and S. P. Muller, Cortical activation to auditory mismatch elicited by frequency deviant and complex novel sounds: A PET study, *NeuroImage*, vol. 17, pp. 231–239, 2002.
3. M. Jing and S. Sanei, Scanner artifact removal in simultaneous EEG–fMRI for epileptic seizure prediction, in *IEEE 18th International Conference on Pattern Recognition (ICPR)*, Vol. 3, New York: IEEE, pp. 722–725, 2006.
4. C. Aine, M. Huang, J. Stephen, and R. Christopher, Multistart algorithms for MEG empirical data analysis reliably characterize locations and time courses of multiple sources, *NeuroImage*, vol. 12, pp. 159–179, 2000.
5. S. Sanei and J. A. Chambers, *EEG Signal Processing*, Hoboken, NJ: John Wiley & Sons, 2013.
6. K. Wendel et al. EEG/MEG source imaging: Methods, challenges, and open issues, *Computational Intelligence and Neuroscience*, vol. 2009, p. 13, 2009.

7. M. A. Jatoi et al. A survey of methods used for source localization using EEG signals. *Biomedical Signal Processing and Control*, vol. 11, pp. 42–52, 2014.

8. M. Teplan, Fundamentals of EEG measurements, *Measurement Neuroscience Reviews*, vol. 2(2), pp. 1–11, 2002.

9. J. H. Margerison, C. D. Binnie, and I. R. McCaul, Electroencephalographic signs employed in the location of ruptured intracranial arterial aneurysms, *Electroencephalography and Clinical Neurophysiology*, vol. 28, pp. 296–306, 1970.

10. SCCN. Available: http://www.sccn.ucsd.edu Accessed on January 16, 2017.

11. H. H. Jasper, The ten twenty electrode system of the international federation, *Electroencephalography and Clinical Neurophysiology*, vol. 10, pp. 371–375, 1958.

12. H. Jasper, Report of committee on methods of clinical exam in EEG, *Electroencephalography and Clinical Neurophysiology*, vol. 10, pp. 370–375, 1958.

13. G. H. Klem et al. The ten-twenty electrode system of the International Federation, *Electroencephalography and Clinical Neurophysiology*, vol. 52(3), pp. 3–6, 1999.

14. Effects of electrode placement. Available: http://www.focused-technology.com/electrod.htm. Accessed on January 16, 2017.

15. T. Collura, *A Guide to Electrode Selection, Location, and Application for EEG Biofeedback*, Bedford, OH: Brain-Master Technologies, Inc., 1998.

16. H. H. Jasper and H. L. Andrews, Electro-encephalography: III. Normal differentiation of occipital and precentral regions in man, *Archives of Neurology and Psychiatry*, vol. 39(1), pp. 96–115, 1938.

17. W. G. Walter, The location of cerebral tumours by electro-encephalography, *The Lancet*, vol. 228(5893), pp. 305–308, 1936.

18. M. B. Sterman, L. R. MacDonald, and R. Kc Stone, Biofeedback training of the sensorimotor electroencephalogram rhythm in man: Effects on epilepsy, *Epilepsia*, vol. 15(3), pp. 395–416, 1974.

19. P. L. Silbert, K. Radhakrishnan, J. Johnson, and D. W. Class, The significance of the phi rhythm, *Electroencephalography and Clinical Neurophysiology*, vol. 95, pp. 71–76, 1995.

20. W. A. Cobb, R. J. Guiloff, and J. Cast, Breach rhythm: The EEG related to skull defects, *Electroencephalography and Clinical Neurophysiology*, vol. 47, pp. 251–271, 1979.

21. E. Roldan, V. Lepicovska, C. Dostalek, and L. Hrudova, Mu-like EEG rhythm generation in the course of Hatha-Yogi exercises, *Electroencephalography and Clinical Neurophysiology*, vol. 52, p. 13, 1981.

22. IFSECN, A glossary of terms commonly used by clinical electroencephalographers, *Electroencephalography and Clinical Neurophysiology*, vol. 37, pp. 538–548, 1974.

23. M. H. Soomro, N. Badruddin, M. Z. Yusoff, and A. S. Malik, A method for automatic removal of eye blink artifacts from EEG based on EMD-ICA, in *2013 IEEE 9th International Colloquium on Signal Processing and Its Applications*, Kuala Lumpur, Malaysia. New York: IEEE, pp. 129–134, 2013.

24. R. Croft and R. Barry, Removal of ocular artifact from the EEG: A review, *Neurophysiologie Clinique/Clinical Neurophysiology*, vol. 30, pp. 5–19, 2000.

25. G. Gratton, Dealing with artefacts: The EOG contamination of the event-related brain potentials over the scalp, *Electroencephalography and Clinical Neurophysiology*, vol. 27, p. 546, 1969.

26. L. Sun and H. Hinrichs, Simultaneously recorded EEG–fMRI: Removal of gradient artifacts by subtraction of head movement related average artifact waveforms, *Human Brain Mapping*, vol. 30, pp. 3361–3377, 2009.

27. R. Niazy, C. Beckmann, G. Iannetti, J. Brady, and S. Smith, Removal of FMRI environment artifacts from EEG data using optimal basis sets, *Neuroimage*, vol. 28, pp. 720–737, 2005.

28. G. Pfurtscheller and F. Lopes da Silva, EEG event-related desynchronization (ERD) and event-related synchronization (ERS), in *Electroencephalography: Basic Principles, Clinical Applications and Related Fields*, Philadelphia, PA: Lippincott Williams & Wilkins, vol. 958, 1999.

29. A. Roman-Gonzalez, EEG signal processing for BCI applications, in *Human–Computer Systems Interaction: Backgrounds and Applications 2*, Berlin, Germany: Springer, pp. 571–591, 2012.

30. T. Thompson, T. Steffert, T. Ros, J. Leach, and J. Gruzelier, EEG applications for sport and performance, *Methods*, vol. 45, pp. 279–288, 2008.

31. K. Awada et al. Effect of conductivity uncertainties and modeling errors on EEG source localization using a 2-D model, *IEEE Transactions on Biomedical Engineering*, vol. 45, pp. 1135–1145, 1998.

32. J. Ebersole, EEG voltage topography and dipole source modeling of epi-leptiform potentials, *Current Practice of Clinical Electroencephalography*, Philadelphia, PA: Lippincott Williams & Wilkins, pp. 732–752, 2003.

33. C. Plummer, A. S. Harvey, and M. Cook, EEG source localization in focal epilepsy: Where are we now?, *Epilepsia*, vol. 49, pp. 201–218, 2008.

34. M. A. Brazier, A study of the electrical fields at the surface of the head, *Electroencephalography and Clinical Neurophysiology*, vol. 2, pp. 38–52, 1949.

35. J. C. Shaw and M. Roth, Potential distribution analysis II. A theoreti-cal consideration of its significance in terms of electrical field theory, *Electroencephalography and Clinical Neurophysiology*, vol. 7, pp. 285–292, 1955.

36. M. R. Schneider, A multistage process for computing virtual dipolar sources of EEG discharges from surface information, *IEEE Transactions on Biomedical Engineering*, vol. 19, pp. 1–12, 1972.

37. C. Henderson, S. Butler, and A. Glass, The localization of equiva-lent dipoles of EEG sources by the application of electrical field theory, *Electroencephalography and Clinical Neurophysiology*, vol. 39, pp. 117–130, 1975.

38. A. Tarantola, *Inverse Problem Theory and Methods for Model Parameter Estimation*, Philadelphia, PA: Society for Industrial and Applied Mathematics, 2005.

39. M. S. Zhdanov, *Geophysical Inverse Theory and Regularization Problems*, Vol. 36, Oxford, UK: Elsevier, 2002.

chapter three

EEG forward problem I
Mathematical background

Introduction

This chapter discusses the mathematical background for the forward problem. The forward problem is concerned with the computation of scalp potentials and external fields for a specific set of neural current sources. Hence, it needs to model the head that is used for calculation of scalp potentials. The head modeling is carried out with various numerical techniques [1,2]. However, prior to this, it is necessary to understand the mathematical background behind it. Thus, this chapter is dedicated to developing an understanding of the mathematical derivations that define the voltage and current relationships for electromagnetic activity, which causes the neural currents to flow and thus activate the brain sources. To achieve this purpose, the chapter starts with the basic explanation of Maxwell's equations, which are governing equations for electromagnetic phenomenon for any medium. Then, it moves to define the quasi-static approximation of Maxwell's equations for the inverse problem. Furthermore, the potential difference that is derived for the forward problem is explained with necessary derivations and formulations. The dipole approximation and conductivity estimation are covered, and finally a summary is provided.

3.1 Maxwell's equations in EEG inverse problems

As explained previously, to solve the EEG forward problem, it is essential to understand Maxwell's equations. Therefore, this section explains the mathematical formulation for all equations of Maxwell, which are termed as Maxwell's equations for electromagnetics, as they define the relationship between electrical and magnetic quantities in terms of time-varying derivations.

The interrelation between various electrical and magnetic quantities was developed by James Maxwell in 1861. The set of equations that was later termed *Maxwell's equations* provides the relationship between electromagnetic field and the charge density and current density. These equations are expressed in both point (differential form) and integral form.

These equations are used to study the time-varying behavior of electric and magnetic fields in various media. The properties of a medium such as permittivity, permeability, conductivity, and impedance strongly affect the equations as we see in the following section.

The set of Maxwell's equations in differential and integral forms for any medium can be given as follows:

Differential form

$$\nabla \times \mathbf{H} = \frac{\partial D}{\partial t} + \mathbf{J} \quad \text{(Ampere's law)} \tag{3.1}$$

$$\nabla \times \mathbf{E} = -\frac{\partial \mathbf{B}}{\partial t} \quad \text{(Faraday's law)} \tag{3.2}$$

$$\nabla \cdot \mathbf{D} = \rho_v \quad \text{(Gauss' law)} \tag{3.3}$$

$$\nabla \cdot \mathbf{B} = 0 \quad \text{(Gauss' law for magnetism)} \tag{3.4}$$

Integral form

$$\oint_L \mathbf{H} \cdot d\mathbf{L} = \int_S \left(\frac{\partial D}{\partial t} + \mathbf{J} \right) dS \tag{3.5}$$

$$\oint_L \mathbf{E} \cdot d\mathbf{L} = -\int_S \frac{\partial \mathbf{B}}{\partial t} \cdot dS \tag{3.6}$$

$$\oint_S \mathbf{D} \cdot dS = \oint_v \rho_v \, dv \tag{3.7}$$

$$\oint_S \mathbf{B} \cdot dS = 0 \tag{3.8}$$

where $\mathbf{H}$ = magnetic field strength (A/m), $\mathbf{E}$ = electric field strength (V/m), $\mathbf{D}$ = electric flux density (C/m²), $\partial D/\partial t$ = displacement electric current density (A/m²), $\mathbf{J}$ = conduction current density (A/m²), $\mathbf{B}$ = magnetic flux density (wb/m² or Tesla), $\partial B/\partial t$ = time-derivative of magnetic flux density (wb/m² s), ρ_v = Volume charge density (C/m³), dL = differential length, dS = differential area, and dv = differential volume.

These equations can be explained in the context of a varying electric or magnetic field. For example, Equation 3.1, which is also Ampere's law, states that the changing magnetomotive force around a closed path will result in the summation of electric displacement and conduction currents through any surface bounded by the path. The second equation, which is also called Faraday's law, states that the electromotive force around a closed path is equal to the negative time derivative of magnetic flux flowing through any surface bounded by the path. Maxwell's third equation, also known as Gauss' law, defines the relationship between electric displacement flux and total charge inside that surface. The last equation is known as Gauss' law for magnetism, which states that the total magnetic flux passing through any closed surface is equal to zero [3–6].

These equations have different versions for different (homogeneous, nonhomogeneous, isotropic, anisotropic, and source-free region) media as the medium permittivity (ε), permeability (μ), and conductivity (σ) change significantly for various mediums. For example, for the free space where relative permittivity $\varepsilon_r = 1$, relative permeability $\mu_r = 1$, conduction current density $J = 0$, and conductivity is zero, the set of Maxwell's equations is given as follows [7–10]:

$$\nabla \times \mathbf{H} = \frac{\partial D}{\partial t} \quad \leftrightarrow \quad \oint_L \mathbf{H} \cdot dL = \int_S \frac{\partial D}{\partial t} dS \tag{3.9}$$

$$\nabla \times \mathbf{E} = -\frac{\partial B}{\partial t} \quad \leftrightarrow \quad \oint_L \mathbf{E} \cdot dL = -\int_S \frac{\partial B}{\partial t} \cdot dS \tag{3.10}$$

$$\nabla \cdot \mathbf{D} = 0 \quad \leftrightarrow \quad \oint_S \mathbf{D} \cdot dS = \oint_v \rho_v \, dv \tag{3.11}$$

$$\nabla \cdot \mathbf{B} = 0 \quad \leftrightarrow \quad \oint_S \mathbf{B} \cdot dS = 0 \tag{3.12}$$

Human tissue exhibits nonhomogeneous characteristics with different values of permittivity ε, permeability (μ), and conductivity (σ) throughout the medium. Hence, the permeability value is the same for vacuum, whereas the relative permittivity value changes as it is dependent on tissue and frequency. According to the literature [11], at the frequency of 100 Hz, $\varepsilon_r = 4 \times 10^6$ for gray matter, 5×10^5 for fat, and 6×10^3 for bone material.

3.2 Quasi-static approximation for head modeling

The frequency range that is important to neuroscientists for brain study
is typically below 1 kHz, and thus most studies deal with frequencies
between 0.1 and 100 Hz. Therefore, the physics related to magnetoen-
cephalography/EEG is well defined using quasi-static approximations
for Maxwell's equations. This approximation is also valid as the brain
signals are generated at lower frequencies only. At these lower frequencies, the time-derivative component in Maxwell's equations is ignored.
Such an approximation for Maxwell's equations is called quasi-static
approximation. This approximation is derived and proved by various
literature surveys such as in Hämäläinen and Sarvas [12]. Therefore,
with the assumption that head tissues are having permeability of free
space, Maxwell's equations now become

$$\nabla \times \mathbf{B} = \mu_0 \mathbf{J} + \mu_0 \varepsilon_r \varepsilon_0 \frac{\partial E}{\partial t} \tag{3.13}$$

$$\nabla \times \mathbf{E} = -\frac{\partial B}{\partial t} \tag{3.14}$$

$$\nabla \cdot \mathbf{E} = \frac{\rho}{\varepsilon_r \varepsilon_0} \tag{3.15}$$

$$\nabla \cdot \mathbf{B} = 0 \tag{3.16}$$

Now using Ohm's law, $\mathbf{J} = \sigma \mathbf{E}$, so Equation 3.13 becomes

$$\nabla \times \mathbf{B} = \mu_0 \left(\sigma \mathbf{E} + \varepsilon_r \varepsilon_0 \frac{\partial E}{\partial t} \right) \tag{3.17}$$

By taking $\mathbf{E}(t) = \mathbf{E}_0 \cdot e^{-j\omega t}$ in Equation 3.17,

$$\nabla \times \mathbf{B} = \mu_0 (\sigma \mathbf{E} - j\varepsilon_0 \varepsilon_r \ \omega \mathbf{E}) \tag{3.18}$$

Thus, for a quasi-static approximation, the term $|\varepsilon_0 \varepsilon_r / \sigma| \ll 1$.
According to the literature [13], the tissue conductivity for the head region
at a frequency of 100 Hz is 0.3 Ω^{-1} m^{-1} with relative permittivity $\varepsilon_r = 10^5$.
Therefore, we get $|\varepsilon_0 \varepsilon_r / \sigma| \ll 1$. A similar analogy can be derived for the
second equation $\nabla \times \mathbf{E} = -\partial B / \partial t$, in which we can ignore the temporal
derivative of magnetic field strength (**B**) because of low frequencies.
Hence, we can write:

$$\nabla \times \mathbf{E} = 0 \tag{3.19}$$

The use of quasi-static assumptions is evident as it makes mathematics analysis simpler by separating both the magnetic and the electric field calculations. Second, the propagation delay of the neural signals transmitted by the active sources inside the brain to the electrodes on the scalp can be ignored. Hence, it can be assumed that the electrode placed on the scalp surface measures the activity of the brain within a particular instant without any delay [14].

3.3 Potential derivation for the forward problem

From this discussion related to Maxwell's equation with quasi-static assumption, we concluded that the curl of the electric field strength is zero. Hence, the gradient of the scalar potential V can be related to field strength as

$$\mathbf{E} = -\nabla \mathbf{V} \tag{3.20}$$

The total current density ($\mathbf{J}$) in a volume with current generators can be categorized into either primary current flow denoted by $\mathbf{J}^p$ or volume currents $\mathbf{J}^v$. Using the definition related to brain volume, the $\mathbf{J}^p$ is the current density due to the neuronal activity in the brain. This current density is spatially bounded in a volume. However, the volume current density ($\mathbf{J}^v$) flows due to the electric field in the volume under observation.

Therefore, using the point form of Ohm's law, $\mathbf{J}^v = \sigma \mathbf{E}$, the total current density is given as follows:

$$\mathbf{J} = \mathbf{J}^p + \mathbf{J}^v = \mathbf{J}^p + \sigma \mathbf{E} = \mathbf{J}^p - \sigma \nabla \mathbf{V} \tag{3.21}$$

In physiological terms concerning the brain, the intracellular currents are primary currents or the impressed currents, and the volume currents are extracellular currents. In addition, the volume currents are produced due to the displacement of charges, which is caused by the gradient of potential in that medium.

For the quasi-static approximation of ignoring the temporal differentiation, Equation 3.18 is modified as follows:

$$\nabla \times \mathbf{B} = \mu_0 \sigma \mathbf{E} = \mu \mathbf{J} \tag{3.22}$$

Taking the gradient on both sides of Equation 3.22, we have

$$\nabla \cdot (\nabla \times \mathbf{B}) = \nabla \cdot (\mu \mathbf{J}) \tag{3.23}$$

$$\Rightarrow \nabla \cdot (\mu \mathbf{J}) = 0$$
$$\Rightarrow \nabla \cdot \mathbf{J} = 0 \tag{3.24}$$

Now by applying a divergence operator on both sides, we have

$$\nabla \cdot \mathbf{J} = \nabla \cdot \mathbf{J}^p - \nabla \cdot (\sigma \nabla \mathbf{V}) \tag{3.25}$$

$$0 = \nabla \cdot \mathbf{J}^p - \nabla \cdot (\sigma \nabla \mathbf{V})$$

$$\nabla \cdot (\sigma \nabla \mathbf{V}) = \nabla \cdot \mathbf{J}^p \tag{3.26}$$

This equation is known as Poisson's equation for the electric potential (**V**) inside an isotropic medium with arbitrary conductivity and insulated boundary. In these equations, **J** is current density with units A/m², and it is a vector field in three dimensions—that is, J(x, y, z). However, taking the divergence of this vector field will generate current source density, which is represented by $\mathbf{I}_m$ and has units of A/m³. Hence,

$$\nabla \cdot (\sigma \nabla \mathbf{V}) = \nabla \cdot \mathbf{J}^p = \mathbf{I}_m \tag{3.27}$$

Poisson's equation defined in Equation 3.27 is the forward problem equation for EEG source localization. The solutions are to be found for this equation to solve the EEG forward problem, where the electric potential satisfies the equation at every location in the head volume conductor model. Hence, with a given head model and current source, the aforementioned equation can be used to evaluate the potential and formulate the forward problem with effective results [15].

For the specialized case where there is no free charge, Poisson's equation is written as follows:

$$\nabla \cdot (\sigma \nabla \mathbf{V}) = 0 \tag{3.28}$$

This equation is known as the Laplace equation for EEG source analysis. As with Poisson's equation, the Laplace equation is useful to calculate the electric potential in regions (like the human head in EEG source analysis) at which the boundary voltage is known.

3.3.1 Boundary conditions

To solve the forward problem for EEG source localization, defined by Equation 3.28 given above, certain boundary conditions should be satisfied. There are two boundary conditions that are defined for different regions of the head during head modeling [16]. They are termed *Neumann* and *Dirichlet* boundary conditions, which are considered at the interfaces between two different regions. The difference between two regions is checked by their respective conductivities (σ_1 and σ_2) and the unit normal vector n' to the interface between the regions.

The Neumann boundary condition states that the charges are not accumulated on the interfaces. Rather they are traveling after leaving one interface. In other words, because the head is a pure resistive medium, there is continuity in current from one interface to the other. Hence, all the current leaving a region with conductivity σ_1 through the interface enters into a neighboring region with conductivity σ_2. Mathematically,

$$\mathbf{J}_1 \cdot n' = \mathbf{J}_2 \cdot n' \tag{3.29}$$

$$(\sigma_1 \nabla \mathbf{V}_1) \cdot n' = (\sigma_2 \nabla \mathbf{V}_2) \cdot n' \tag{3.30}$$

where n' is the normal vector on the interface.

A special case of the Neumann boundary condition is for a homogeneous medium, which is called the homogeneous Neumann boundary condition [17]. This states that at the outer surface of the human head, no current can flow from the head into air due to the low conductivity of air. Hence, mathematically it can be defined as follows:

$$\mathbf{J}_1 \cdot n' = 0 \tag{3.31}$$

$$(\sigma_1 \nabla \mathbf{V}_1) \cdot n' = 0 \tag{3.32}$$

The other boundary condition for the forward problem solution is the Dirichlet boundary condition. It is limited to internal interfaces and explains the potential at the boundary; hence, it states that the electric potential shows continuity across the interfaces, such that

$$\mathbf{V}_1 = \mathbf{V}_2 \tag{3.33}$$

This equation represents the Dirichlet boundary condition. Besides this, a reference electrode with zero potential is also assigned such that

$$\mathbf{V}_{ref} = 0 \tag{3.34}$$

Hence, for the forward problem solution with different head models, which are to be used for the calculation of potential due to dipole sources, the aforementioned boundary conditions are taken into consideration. This will lead to a proper solution with less errors and more resolution.

3.4 Dipole approximation and conductivity estimation

According to the literature available for bioelectromagnetism [18–20], the electrical activity inside the brain can be modeled as a current dipole. This dipole provides good approximation for a small source viewed from a relatively large distance, as the far field of a realistic source is mainly dipolar. This approximation is valid to model the electrical activity in the active region of the cerebral cortex at a single time instant. EEG detects the activity of the pyramidal cells, which are aligned parallel to each other and perpendicular to the cortical surface. Because of the large distance between EEG electrodes and the active region area, the electrical activity measured has dipolar configuration; that is, scalp voltage topography has two maxima, one negative and one positive [21]. Hence, an equivalent current dipole can be used for modeling the electrical activity of the cortex.

As shown in Figure 3.1, the current dipole consists of the current source and the sink separated by distance p with an equal amount of current to be injected or removed. The magnitude and orientation of dipole are characterized by the dipole moment d vector, which is pointing from the source to the sink in the figure. Hence, the magnitude of the dipole is $|d| = Ip$, where p is the distance between the positive and negative poles, respectively.

Vector notation for the dipole moment can be written as:

$$d = \mathbf{I}pn_d'$$ (3.35)

where n_d' is the unit vector defining the direction from the source to the sink. However, in the Cartesian coordinate system, the dipole can be represented by

$$d = d_x x^\wedge + d_y y^\wedge + d_z z^\wedge$$ (3.36)

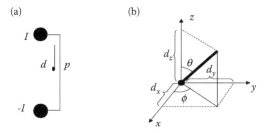

Figure 3.1 (a) Current dipole with distance and (b) Cartesian coordinates representation.

where $x^\wedge$, $y^\wedge$, and $z^\wedge$ are basis vectors with unit magnitude for the Cartesian coordinate system, with d_x, d_y, and d_z as Cartesian components of the dipole.

Because Poisson's equation is linear, the superposition rule can be applied to separate dipoles to generate a summation result for the net dipole. This principle of superposition is more important when we are dealing with the spatiotemporal dipole model as the single dipole activity can be summed up to represent a complex current source over a span of time. In this way, a potential **V** at an arbitrary point, generated by a dipole at a position r and orientation d, can be written in Cartesian coordinates as:

$$\mathbf{V}(r,d) = d_x\mathbf{V}(r,n_x) + d_y\mathbf{V}(r,n_y) + d_z\mathbf{V}(r,n_z) \tag{3.37}$$

Tissue conductivity is an important parameter for EEG source analysis when formulating head conductor models. Different models use different regions, such as the scalp, skull, and brain or cerebrospinal fluid, gray matter, and white matter. According to the literature, the source analysis is largely affected by the incorrect values for tissue conductivity. Some noninvasive techniques are used for measuring the conductivity of the head such as diffusion tensor imaging or electrical impedance tomography. However, it is assumed mostly that the scalp and brain have similar conductivities, whereas the skull has relatively smaller conductivity. The commonly used scalp-to-skull/scalp-to-brain conductivity ratio is 1:1/80:1, reported in Rush and Driscoll [22]. However, the latest developments for *in vivo* and *in vitro* conductivity measurements suggest the conductivity ratio of 1:1/5:1 [23]. Further studies are provided [13] that are related to the impact of soft bone on EEG recordings.

Summary

This chapter provides a foundation to explain the mathematical formulation being employed for solving the forward problem. Thus, it started with an explanation of Maxwell's equations, which are considered as basic equations to understand any electromagnetic phenomenon. Moreover, the mathematical assumption applied for brain signals was covered in the "Quasi-Static Approximation for Head Modeling" section. The dipole, which is considered as equivalent to the brain source, was defined as explained with the help of proper formulations. However, the conductivity values for various brain regions were explained according to those provided in the literature. Hence, this chapter gives a brief and precise mathematical background for the forward problem.

References

1. J. Kybic et al. A common formalism for the integral formulations of the forward EEG problem, *IEEE Transactions on Medical Imaging*, vol. 24, pp. 12–28, 2005.
2. H. Hallez et al., Review on solving the forward problem in EEG source analysis, *Journal of Neuroengineering and Rehabilitation*, vol. 4, p. 1, 2007.
3. A. Taflove and S. C. Hagness, *Computational Electrodynamics*, Norwood, MA: Artech House, 2005.
4. K. Yee, Numerical solution of initial boundary value problems involving Maxwell's equations in isotropic media, *IEEE Transactions on Antennas and Propagation*, vol. 14, pp. 302–307, 1966.
5. T. Weiland, A discretization model for the solution of Maxwell's equations for six-component fields. *Archiv Elektronik und Uebertragungstechnik*, vol. 31, pp. 116–120, 1977.
6. K. S. Kunz and L. Simpson, A technique for increasing the resolution of finite-difference solutions of the Maxwell equation, *IEEE Transactions on Electromagnetic Compatibility*, vol. 4, pp. 419–422, 1981.
7. C. T. A. Johnk, *Engineering Electromagnetic Fields and Waves*, Vol. 667, New York: John Wiley & Sons, p. 1, 1975.
8. H. A. Haus and J. R. Melcher, *Electromagnetic Fields and Energy*, Upper Saddle River, NJ: Prentice Hall, 1989.
9. W. C. Chew, *Waves and Fields in Inhomogeneous Media*, Vol. 522, New York: IEEE Press, 1995.
10. G. T. Markov and A. F. Chaplin, *The Excitation of Electromagnetic Waves*, Moscow, Russia: Moscow Izdatel Radio Sviaz, 1983.
11. S. Gabriel, R. Lau, and C. Gabriel, The dielectric properties of biological tissues: II. Measurements in the frequency range 10 Hz to 20 GHz, *Physics in Medicine and Biology*, vol. 41, p. 2251, 1996.
12. M. S. Hämäläinen and J. Sarvas, Realistic conductivity geometry model of the human head for interpretation of neuromagnetic data, *IEEE Transactions on Biomedical Engineering*, vol. 36, pp. 165–171, 1989.
13. C. Ramon, P. Schimpf, J. Haueisen, M. Holmes, and A. Ishimaru, Role of soft bone, CSF and gray matter in EEG simulations, *Brain Topography*, vol. 16, pp. 245–248, 2004.
14. G. Pruis, B. H. Gilding, and M. Peters, A comparison of different numerical methods for solving the forward problem in EEG and MEG, *Physiological Measurement*, vol. 14, p. A1, 1993.
15. K. A. Awada et al., Effect of conductivity uncertainties and modeling errors on EEG source localization using a 2-D model, *IEEE Transactions on Biomedical Engineering*, vol. 45, pp. 1135–1145, 1998.
16. K. Awada and R. Greenblatt, EEG source localization conductivity sensitivity using a 3-D finite element model, *Neuroimage*, vol. 9, p. S138, 1999.
17. K. A. Awada et al., Closed-form evaluation of flux integrals appearing in a FEM solution of the 2D Poisson equation with dipole sources, *Electromagnetics*, vol. 16, pp. 75–90, 1996.
18. J. C. De Munck, B. W. Van Dijk, and H. Spekreijse, Mathematical dipoles are adequate to describe realistic generators of human brain activity, *IEEE Transactions on Biomedical Engineering*, vol. 35, pp. 960–966, 1988.

19. J. Sarvas, Basic mathematical and electromagnetic concepts of the biomagnetic inverse problem, *Physics in Medicine and Biology*, vol. 32, p. 11, 1987.

20. J. S. Ebersole, EEG voltage topography and dipole source modeling of epileptifor potentials, *Current Practice of Clinical Electroencephalography*, Philadelphia, Lippincott Williams & Wilkins, 2003, pp. 732–752.

21. J. S. Ebersole, Noninvasive localization of epileptogenic foci by EEG source modeling, *Epilepsia*, vol. 41, pp. S24–S33, 2000.

22. S. Rush and D. A. Driscoll, Current distribution in the brain from surface electrodes, *Anesthesia & Analgesia*, vol. 47, pp. 717–723, 1968.

23. M. Akhtari et al., Conductivities of three-layer live human skull, *Brain Topography*, vol. 14, pp. 151–167, 2002.

chapter four

EEG forward problem II
Head modeling approaches

Introduction

This chapter continues the discussion pertaining to the forward problem by explaining the head modeling schemes, which are used to model the head. These head modeling schemes are based on mathematical formulations, which are either analytical in nature or numerical depending on particular applications. Hence, according to the discussion in the last chapter related to tissue conductivity, it is clear that the human brain is neither a homogeneous nor an infinite conductor. Rather, it is a type of bounded conductor as the electrical current is unable to flow outside it. In addition, the layers of the brain such as the skull and scalp have variations in conductivity. This inhomogeneity must be considered when head modeling is carried out for solving the forward problem. Hence, specific to application, accuracy, and computational burden, a volume conductor model can be designed for the solution of the forward problem. In addition, the head geometry and tissue inhomogeneity are the main concerns when developing a volume conductor model. With the simplest assumption, the head is assumed as a set of nested concentric spheres. With this simple geometry having constant isotropic conductivity, the electrical potential can be calculated analytically for the forward problem. However, the simpler geometry leads to a rough approximation with more localization error [1–3].

In general, the volume conductor head models are classified as analytical models and numerical models, respectively. The analytical solution uses spherical head modeling with homogeneous conductivity for each sphere [4]. Hence, the electric potential is calculated using such a simple geometrical model and conductivity distribution. The numerical models, however, are provided with much more realistic modeling of the human head with more complexity but with more freedom as compared with analytical models. Because of complicated geometries and conductivity distribution, the computational complexity of numerical models is high but yields an accurate solution with high resolution and low localization error. The most commonly applied numerical head modeling schemes for the forward problem are the boundary element method (BEM), the finite difference method (FDM), and the finite element method (FEM). These models with necessary mathematical formulations are explained in the following sections.

4.1 Analytical methods versus numerical methods for head modeling

In this section, a brief discussion is provided for analytical head models and numerical head models, respectively.

4.1.1 Analytical head modeling

This is the simplest head modeling scheme, which takes into account the geometry of the head as nested concentric spheres. These spheres are treated as volume conductors to represent layers of the human head. In the literature, these layers were first reported with a homogeneous sphere [4]. However, due to the larger difference between the conductivity of skull and scalp/brain tissue, the model was refined to a three-shell concentric model. For such a model, the semianalytic solution of Poisson's equation was proposed [5]. In some studies, a five-shell model is also found to represent more tissue types for the human head. However, the analytical expressions representing the spherical models are different depending on the exact configuration of the volume conductor, and most of the time they are represented by an infinite series of Legendre polynomials for the solution of Poisson's equation [6,7]. The calculation of the potentials with the assumption of nested shells with certain radii and conductivities with number of electrodes is available in the literature. These analytical methods allow us to perform calculations of electric potential with the constraint of dividing the human head into nested spheres with constant conductivities.

The studies showing comparison between realistic head models and spherical head models [8,9] show that due to high simplicity and ignoring of important parameters, the spherical models are subjected to a high number of errors and less accurate EEG source localization. In some research articles, spherical head models were reported to have dipole localization error in several centimeters, compared with realistic head models.

The mathematical formulation employed for analytical models can be extended for nonconcentric spheres or ellipsoidal geometries [10]. However, most of the time, the simplicity of such models leads to less reliable results. Hence, it is mostly recommended by researchers to use realistic head models with numerical solutions to approximate properly the active sources using EEG signals. However, some researchers use the spherical models for the solution of the forward problem just to check the accuracy of the numerical head volume conductor models. Therefore, we can say that the analytical spherical models are used as the "gold standard" by which the performance capability of numerical models is analyzed for source localization.

4.1.2 Numerical head models

The numerical head volume conductor models are used to increase the resolution and to improve the localization capability of EEG source localization. These models are based on the assumptions of realistic head models, which are complex in geometry, but this results in accurate localization. The geometry of these models is obtained by anatomical imaging modalities such as computed tomography and structural magnetic resonance imaging. This phenomenon of taking three-dimensional (3D) anatomical data results in complex geometry and hence increased complexity. The realistic head model is realized using MR images to obtain anatomical information, which includes head shape and tissue distributions. Normally, the image is segmented using the standard segmentation procedure to divide it into several regions having the same voxel intensity.

The increased computational capabilities offered by the latest technologies have made it convenient to manipulate complex mathematical models involved in the numerical solution of Poisson's equation. Hence, the numerical head volume conductors apply various numerical methods to calculate an approximate solution of Poisson's equation. Their ability to model the head with realistic parameters using various tissue conductivity distributions makes them more efficient and versatile as compared with analytical models. Furthermore, these models are equally valid for nonhomogeneous and anisotropic conductivity distributions.

The numerical methods involve the generation of mesh for the domain (in our case the human head) for the solution of Poisson's equation. This step will generate the small patches in which the physical quantity (potential) can be calculated. This partitioning is based on the tissue type and relative conductivity of the region (scalp, skull, cerebrospinal fluid, and brain). The computational grid is then assigned various nodes. After this, appropriate boundary conditions (such as Neumann and Dirichlet) are applied [11].

This procedure results in a system of algebraic equations expressed in matrix form:

$$Ax = b \tag{4.1}$$

where A is a $\Re^{N \times N}$ matrix, and x is a $\Re^{N \times 1}$ vector. The elements of x are electric potentials, and b is a $\Re^{N \times 1}$ vector having current source values. At the end, Poisson's equation is solved using numerical methods to estimate the values of electric potentials. The inversion of such a matrix is done using standard inversion techniques. In this way, Poisson's equation for realistic head models is solved using numerical techniques. A detailed discussion for numerical methods (BEM, FDM, and FEM) is provided in the following sections.

4.2 Finite difference method

The FDM is the commonly applied numerical method for the solution of partial differential equations. FDM solves differential equations by the approximation of differentials with finite differences—that is, approximate equivalent difference quotients. This method is used to solve the field problems for various applications, which include heat transfer, fluid mechanics, biomechanics, and bioelectromagnetism. Thus, the head models for the EEG forward problem are also constructed using FDM to localize the brain sources. FDM models the entire head volume by partitioning it into a regular grid of hexahedral volume elements [12]. Each element is assigned its own conductivity based on the tissue type. The mapping between the MR image and FDM voxels is straightforward as the voxels generated in the mesh process of FDM represent the positions and size in the MR image. The partial derivatives with truncated Taylor series expansions are substituted to have a discretized version of Poisson's equation. Hence, if the function $f(x)$ has a first-order derivative as:

$$f'(x) = \lim_{h \to 0} \frac{f(x+h) - f(x)}{h} \tag{4.2}$$

Then for Poisson's equation $\nabla \cdot (\sigma \nabla V)$ in 3D to be approximated by differential operator

$$\nabla \cdot (\sigma \nabla V) \approx \frac{1}{r_0 h^2} \left[\alpha_0 V(r_0) - \sum_{i=1}^{6} \alpha_i V(r_i) \right] \tag{4.3}$$

where α_0 and α_i are the conductivities at the points r_0 and r_i, respectively. From the aforementioned equation, it can be deduced that the electric potential at node 0 having six neighboring nodes can be written as:

$$\left(\sum_{i=1}^{6} \alpha_i \right) V_0 - \sum_{i=1}^{6} \alpha_i V_i = \nabla \cdot \mathbf{J}^\mathrm{p} = \mathbf{I} \tag{4.4}$$

where V_i is electric potential at node i, $\mathbf{I}$ is current leaving the volume element, and α_i is conductance between neighboring nodes expressed as:

$$\alpha_i = \frac{\sigma_0 \sigma_i}{\sigma_0 + \sigma_i} 2h \qquad (4.5)$$

In this way, the FDM is used to solve Poisson's equation related to the EEG forward problem. Various changes are introduced in the FDM model to implement it for anisotropic conductivity [13]. The basic advantage of using FDM as the numerical tool is its coverage for tissue discontinuities, tissue inhomogeneities, and anisotropic conductivity, which are not included in analytical (spherical models) or numerical BEM. In addition, the electric potential is calculated for the entire head volume, which gives more detailed information for the solution of the inverse problem. The basic disadvantage of FDM is a cubic grid due to which the complicated interfaces between brain structures and thin layers are not modeled properly. The FDM can be implemented by either writing source code in MATLAB® or through some software packages such as NETSTATION [14] and Fieldtrip [15].

4.3 Finite element method

FEM is a very useful numerical tool, which is used for solving boundary value problems that are defined by a differential equation with a set of boundary conditions [11]. FEM is utilized for the solution of various field problems, which include stress analysis, biomechanics, heat transfer, fluid flow, biomedical, electromagnetics, and so on. In FEM analysis, the domain under observation is discretized into small subdomains (finite elements), and the unknown quantity inside the element is interpolated based on the value provided at the nodes [16–20]. Hence, FEM goes through seven major steps, which are mesh generation, selection of proper interpolation/basis function, conversion of integral equations into linear equations (i.e., weak formulation), assemblage of the system, imposing of boundary conditions (Dirichlet or Neumann), solution of linear equations using common linear algebra techniques, and finally postprocessing of the results. Along with other field problems, FEM is also used for head modeling for the solution of the forward problem in EEG source analysis. FEM head modeling is the strongest numerical technique as it provides more accurate modeling with complex geometries such as with the human head, which has different tissue boundaries and conductivity distributions [21,22].

The mesh generation principle of FEM resembles FDM in which the head is partitioned into smaller patches or finite elements. However, the geometrical flexibility is provided with FEM as it can work with irregular grids of hexahedrons. Such flexibility provides more accuracy when they are mapped into MR images for head modeling. Keeping in mind the anisotropic and isotropic conductivity, FEM can work in both approaches.

The computational nodes are located on the vertices of the finite elements irrespective of the structure of the grid. For the computation of the potential, two methods are adopted for FEM. One formulation is called a direct method, and another is called a subtraction method. In the direct formulation, the total potential is unknown that is solved for, and the dipole source is directly incorporated into the model. However, in the subtraction form, the unknown is the difference between the total potential and the potential due to the same dipole when it is placed in a region of homogeneous conductivity [23]. The system matrix generated from both of these methods is the same. However, the subtraction method is more computationally complex as it requires the computation of flux integrals at the boundaries of elements involved in the calculations.

For the explanation of these methods, let us start with Poisson's equation with a Neumann boundary condition as no electric current is flowing outside the medium Ω (head). Therefore, we can write:

$$\nabla \cdot (\sigma \nabla V) = \nabla \cdot \mathbf{J}^P = \lim_{l \to 0} \mathbf{I}[\delta(r - r^+) - \delta(r - r^-)] \tag{4.6}$$

Subject to the Neumann boundary condition,

$$\sigma \frac{\partial V}{\partial n} = 0$$

For the direct method, Poisson's equation is multiplied by a basis function ϕ_i, and it is integrated over the whole head domain Ω:

$$\int_\Omega \nabla \cdot (\sigma \nabla V) \cdot \phi_i d\Omega = \int_\Omega \nabla \cdot \mathbf{J}^P \cdot \phi_i d\Omega \tag{4.7}$$

By applying integration by the parts formula, we have

$$\int_\Omega \nabla \cdot \mathbf{J}^P \phi_i d\Omega = \int_{\partial \Omega} \phi_i \sigma \nabla \mathbf{V} \cdot dS - \int_\Omega \sigma \nabla \mathbf{V} \cdot \nabla \phi d\Omega \tag{4.8}$$

Hence, the weak formulation for the forward problem will be

$$\int_\Omega \nabla \phi_i \cdot (\sigma \nabla \mathbf{V}) d\Omega = \int_\Omega \phi_i \mathbf{I} d\Omega \tag{4.9}$$

The aforementioned formulation of the FEM problem is done using Galerkin's method [24]. The other method used for reformulation is the

Rayleigh–Ritz method. Irrespective of methods employed for weak formulation, the next step for FEM is the conversion of the reformulated Poisson's equation into discrete equations. Hence, in the FEM head models, the unknown electric potential over the entire computational domain is approximated by

$$\mathbf{V}(x,y,z) = \sum_{i=1}^{N} \mathbf{V}_i \alpha_i(x,y,z) \qquad (4.10)$$

where N is the number of nodes in the grid; $\mathbf{V}_i$ is the electric potential at node i; and $\alpha_i(x, y, z)$ is the basis function related to node i. The basis function is an interpolation function that has zero value outside of the finite element connected to the corresponding node, and it is a spanning factor for the space of piecewise polynomial functions [25]. The system matrix generated by FEM modeling is sparse because of the contribution of the neighboring node to the electric potential at every node. In addition, the FEM system matrix is symmetric and is positive definite, and thus it can be applied as a fast iterative solver.

FEM has the advantage of flexibility provided in head modeling by taking into consideration the tissue discontinuities, tissue inhomogeneities, and anisotropic conductivity. These features make it far more useful in providing realistic head models with numerical solutions with more accuracy and resolution as compared with FDM. However, the selection of the best current dipole to fit into the FEM model is a major problem in head modeling using FEM. This is due to the irregular grid used in FEM.

Many software packages are available to implement FEM for various fields; for example, ANSYS is available to model mechanical engineering problems [26]. Likewise, for head modeling, which is to be used for EEG source localization, various software packages are used. To name a few, *Fieldtrip, BESA* [27], and *SimBio* [28] are used for FEM modeling.

4.4 Boundary element methods

One of the numerical methods used to solve Poisson's equation in a realistic head volume conductor model is the boundary element method (BEM) [29]. It can be defined as a numerical technique that is used for the solution of linear partial differential equations defined over certain boundaries of various domains after their conversion from a differential to an integral form (weak formulation). The structural information of the subject is taken using MR images. The boundary integral equations are formed from the forward problem. After this, the resultant equations are discretized, which means that the surfaces are divided into triangular domains and hence the unknown parameters are approximated by linear combinations of basis functions. The homogeneous conductivity of each domain is assumed for this formulation. Hence, BEM head models represent the various tissues (having different conductivities) with a series of nested regions characterized by homogeneous isotropic conductivities. This feature resembles spherical head modeling, except for the difference of their realistic shaping. In this way, BEM computes the electric potentials at discrete boundaries between homogeneous isotropic conducting regions.

Depending on the application and keeping in view the complexity, BEM uses various shell models. The tiled surfaces of triangles are generated using structural information from MRIs to approximate accurately the actual shapes of the boundaries between various head regions. In this way, the electric potential is calculated for shell surfaces having computational nodes located at a point on every triangular tile. Hence, keeping in view this approximation, one can model the human head as a domain Ω, which is composed of various subregions. Ω_k is separated by surface S_k having a constant conductivity of σ_k and with zero conductivity outside the domain. Assume the three-region model (brain, skull, and scalp) with three interfaces, which are air–scalp interface, scalp–skull interface, and skull–brain interface, represented by S_1, S_2, and S_3, respectively. From Poisson's equation, an integral equation is derived to calculate the potential V at $r \in S_k$ as:

$$\mathbf{V}(r) = \frac{2\sigma_0}{\sigma_k^- + \sigma_k^+}\mathbf{V}_0(r) + \frac{1}{2\pi}\sum_{j=1}^{3}\frac{\sigma_j^- - \sigma_j^+}{\sigma_k^- + \sigma_k^+}\int_{r'\in S_j}\mathbf{V}(r')\frac{r'-r}{\|r'-r\|^3}dS \quad (4.11)$$

where σ_0 is the medium in which the dipole source is located; $\mathbf{V}_0(r)$ is the electric potential in r for an infinite medium; dS is a vector orthogonal to a surface element, and $\|ds\|$ is the area of that surface element; σ_i^+ is medium conductivity located to the exterior of the interface S_i; and σ_i^- is medium conductivity located to the interior of the interface S_i.

The first term in the aforementioned equations calculates the potential for an infinite medium. However, the second term corrects for applying a bounded medium by introducing sources on the interfaces. The equation stated above is used to calculate the potential after discretization of the interface S_i with triangular meshes. The set of linear equations obtained can be expressed as:

$$\begin{bmatrix} V^1 \\ V^2 \\ V^3 \end{bmatrix} = \begin{bmatrix} V_0^1 \\ V_0^2 \\ V_0^3 \end{bmatrix} + \begin{bmatrix} R^{11} & R^{12} & R^{13} \\ R^{21} & R^{22} & R^{23} \\ R^{31} & R^{32} & R^{33} \end{bmatrix}\begin{bmatrix} V^1 \\ V^2 \\ V^3 \end{bmatrix} \quad (4.12)$$

The $\mathbf{V} = [V^1 \quad V^2 \quad V^3]^T$ are unknown potentials at the center of triangles in the surface, and $\mathbf{V}_0 = [V_0^1 \quad V_0^2 \quad V_0^3]^T$ are the potentials due to dipoles at the centers of triangles in an infinite medium. However, the coefficients of R^{ij} provide the contributions of potentials at the interface j to the potentials at the interface i. These coefficients are calculated by solving the integral part in Equation 4.11. Hence, the solution of the equation is given by

$$\mathbf{V} = (I_n - R)^{-1}\mathbf{V}_0 \tag{4.13}$$

where $\mathbf{I}_n$ is the identity matrix. From Equation 4.13, the potentials at the boundary can be calculated after the calculation of $(\mathbf{I}_n - R)^{-1}$. For the rest of the dipoles, the process is revised as $(\mathbf{I}_n - R)^{-1}$, which is independent of dipole parameters.

BEM generates the head models that are simplest among the other numerical methods (FEM, finite volume method, etc.). The limitations arise due to its assumption of isotropic conductivity within closed surfaces. Because of such assumptions, the important features of head modeling, such as tissue discontinuities, inhomogeneity, and isotropic conductivities, are discarded. Furthermore, BEM head models are not very useful for accurate modeling of the head as a whole. Rather, they are useful for the forward problem in small surfaces only. In addition, because direct methods for inverting the system matrix are used in solving the BEM system matrix, there is a reduction in computational efficiency as compared with iterative techniques.

Many formulations of BEM are proposed in the literature. Some of them are those proposed by Geselowitz [30] and an other is called symmetric BEM. The limitations offered by Geselowitz such as increased numerical error and less accuracy are handled by its latest version *SymBEM* developed by Adde et al. [31]. The software packages available for the solution of the forward problem with the help of BEM head modeling include *Fieldtrip*, *OpenMEEG* [32], *SPM* [33], and *Brainstorm* [34]. They are based on different programming languages such as MATLAB, FORTRAN, C/C++, Python, C/MATLAB. These packages are used to generate a head model in standard BEMs as well as symmetric BEMs.

For the boundary conditions of BEM, Green's theorem is used, which is explained briefly in the following sections.

Green's Theorem for BEM

In general terms, Green's theorem is a vector identity, which is equivalent to the curl theorem in the plane. Hence, over a region D in the plane with boundary, Green's theorem states:

$$\oint_{\partial D} P(x,y)dx + Q(x,y)dy = \iint_D \left(\frac{\partial Q}{\partial x} - \frac{\partial P}{\partial y} \right) dxdy \tag{4.14}$$

where the integral on the left side is line integral, and the integral on the right side is surface integral.

It is applied to solve two-dimensional flow integrals where the summation of fluid flowing out from volume is equal to the total outflow summed about an enclosed area.

Hence, in the case of numerical methods—that is, BEM and FEM—Poisson's equation is solved using Green's theorem [35]. Hence, Equation 4.14 is multiplied by a test function ϕ and then integrated over a volume G, which is the head in this case. The resulting equation will be as follows:

$$\int_G \nabla\phi\cdot(\sigma\nabla\mathbf{V})dG = \int_{\partial G}\phi\sigma\nabla\mathbf{V}\cdot dS - \int_G \phi(\nabla\sigma\nabla\mathbf{V})dG \qquad (4.15)$$

After the application of boundary conditions, the weak formulation for the forward problem using such methods will be as follows:

$$-\int_G \nabla\phi\cdot(\sigma\nabla\mathbf{V})dG = \int_G \phi\mathbf{I}_m dGs \qquad (4.16)$$

Thus, the entire volume is discretized into small elements. However, the computational points $\{V_i\}_{i=1}^n$ are identified through vertices of elements. Here, n is the number of vertices. Thus, the unknown potential $\mathbf{V}(x, y, z)$ is calculated as:

$$\mathbf{V}(x,y,z) = \sum_{i=1}^n \mathbf{V}_i\phi_i(x,y,z) \qquad (4.17)$$

where $\phi_{i=1}^n$ is a set of test functions, which are also termed a basis function.

Green's function is used to obtain smoothing as proposed by Harrison et al. [36]. Here, the function is based on a Laplacian matrix graph solved using vertices and faces provided by the structural MRI, considering the intervoxel distance and connections between the sulci. Hence, Green's function $\mathbf{Q}_G \in \Re^{Nd\times Nd}$ is defined as:

$$\mathbf{Q}_G = e^{\sigma G_L} \qquad (4.18)$$

where σ is the positive constant value, which defines the size of activated regions; and $G_L \in \Re^{Nd\times Nd}$ is a Laplacian matrix graph with interdipole connectivity information.

After the detailed discussion concerning numerical methods used for the solution of the forward problem of EEG source analysis, here is the summary provided as a comparison between all. The basic difference between the three methods is of usage of the domain for the solution calculation. The FDM and FEM propose a solution for the entire head volume, whereas the BEM calculates the solution for boundaries between the homogeneous isotropic conducting regions. In this way, the FEM and FDM have a large system matrix and hence increased computational complexity as compared with BEM. The electric potentials for FEM and FDM are calculated using interpolation functions, whereas for the BEM it is necessary to reapply the numerical integration or Barnard formula.

In terms of computational complexity, BEM is much better than FEM and FDM. It is because BEM is a simple inversion of matrix $(1_n - R)^{-1}$. This inversion can be done by simple straightforward linear algebra techniques such as Gauss–Jordan elimination. However, as explained earlier, FEM and FDM system matrices are very large and sparse in nature. Therefore, these matrices with high dimensions are not easy to invert through direct algebraic methods. Only high-performing computers can reverse them with iterative solvers. Hence, with respect to computational load, BEM is better as compared with FEM and FDM. The fixed nature of computational nodes in FDM makes it difficult to manipulate as compared with FEM and BEM, which provide flexibility in choosing the computational nodes. In this sense, FEM can better represent irregular girds as compared with FDM.

In short, BEM uses surfaces as the domain, whereas FEM and FDM use volume for the computational points. The system matrix for FEM and FDM is sparse, whereas that for BEM is full. The computational complexity for BEM is low, whereas for FEM and FDM it is high. The solution provided with BEM is a direct inversion, whereas with FEM and FDM it is an iterative procedure. Therefore, with the introduction of modern high-technology and computational-efficient systems and iterative solvers, BEM is the best choice to solve the forward problem with more regions, high resolution, and efficient computation of an irregular grid. Hence, for the further implementation phase in this book, we have employed BEM as the numerical head modeling scheme for the solution of the forward problem.

Summary

This chapter serves as a conclusive chapter for the discussion of the forward problem, which was divided into two sections to improve readability and for additional clarity. Hence, the discussion is provided for all techniques, which are normally employed for head modeling. It was seen that although numerical models are more complex, they have more resolution and good performance for source localization problems as compared with analytical methods. Hence, mostly the numerical techniques are applied to solve this problem. These numerical methods are carried out by a certain procedure that involves mesh generation, domain discretization, equation solution, and system assemblage. Hence, commonly FEM, BEM, and FDM are used to model the head for a high-resolution solution of brain source localization. Among them, BEM is simpler as compared with FEM as it is noniterative in nature and has less computational complexity as it uses the surface as the domain rather than volume as in the case of FEM and FDM, respectively. Hence, for most applications where low computational complexity is needed with good resolution, BEM is applied with certain software packages.

References

1. C. Ramon, P. Schimpf, J. Haueisen, M. Holmes, and A. Ishimaru, Role of soft bone, CSF and gray matter in EEG simulations, *Brain Topography*, vol. 16, pp. 245–248, 2004.
2. G. Huiskamp, M. Vroeijenstijn, R. Van Dijk, G. Wieneke, and A. C. Van Huffelen, The need for correct realistic geometry in the inverse EEG problem, *IEEE Transactions on Biomedical Engineering*, vol. 46, pp. 1281–1287, 1999.
3. B. N. Cuffin, Effects of head shape on EEGs and MEGs, *IEEE Transactions on Biomedical Engineering*, vol. 37, pp. 44–52, 1990.
4. E. Frank, Electric potential produced by two point current sources in a homogeneous conducting sphere, *Journal of Applied Physics*, vol. 23, pp. 1225–1228, 1952.
5. Y. Salu, L. G. Cohen, D. Rose, S. Sxato, C. Kufta, and M. Hallett, An improved method for localizing electric brain dipoles, *IEEE Transactions on Biomedical Engineering*, vol. 37, pp. 699–705, 1990.
6. J. C. de Munck, The potential distribution in a layered anisotropic spheroidal volume conductor, *Journal of Applied Physics*, vol. 64, pp. 464–470, 1988.
7. H. Zhou and A. Van Oosterom, Computation of the potential distribution in a four-layer anisotropic concentric spherical volume conductor, *IEEE Transactions on Biomedical Engineering*, vol. 39, pp. 154–158, 1992.
8. B. J. Roth, M. Balish, A. Gorbach, and S. Sato, How well does a three-sphere model predict positions of dipoles in a realistically shaped head? *Electroencephalography and Clinical Neurophysiology*, vol. 87, pp. 175–184, 1993.
9. N. Chauveau, X. Franceries, B. Doyon, B. Rigaud, J. P. Morucci, and P. Celsis, Effects of skull thickness, anisotropy, and inhomogeneity on forward EEG/ERP computations using a spherical three-dimensional resistor mesh model, *Human Brain Mapping*, vol. 21, pp. 86–97, 2004.
10. J. W. Meijs and P. J. Peters, The EEG and MEG, using a model of eccentric spheres to describe the head, *IEEE Transactions on Biomedical Engineering*, vol. 34, pp. 913–920, 1987.
11. M. A. Jatoi, N. Kamel, A. S. Malik, I. Faye, and T. Begum, Representing EEG source localization using finite element method, in *2013 IEEE International Conference on Control System, Computing and Engineering (ICCSCE)*, New York: IEEE, 2013, pp. 168–172.
12. M. A. Jatoi et al., EEG-based brain source localization using visual stimuli, *International Journal of Imaging Systems and Technology*, vol. 26.1, pp. 55–64, 2016.

13. H. I. Saleheen and K. T. Ng, New finite difference formulations for general inhomogeneous anisotropic bioelectric problems, *IEEE Transactions on Biomedical Engineering*, vol. 44, pp. 800–809, 1997.

14. EGI. Available: http://www.egi.com/research-division/geodesic-eeg-system-components/eeg-software. Accessed on January 16, 2017.

15. R. Oostenveld, P. Fries, E. Maris, and J.-M. Schoffelen, FieldTrip: Open source software for advanced analysis of MEG, EEG, and invasive electrophysiological data, *Computational Intelligence and Neuroscience*, vol. 2011, p. 9, 2011.

16. K. A. Awada et al., Computational aspects of finite element modeling in EEG source localization, *IEEE Transactions on Biomedical Engineering*, vol. 44.8, pp. 736–752, 1997.

17. K. A. Awada et al., Effect of conductivity uncertainties and modeling errors on EEG source localization using a 2-D model, *IEEE Transactions on Biomedical Engineering*, vol. 45.9, pp. 1135–1145, 1998.

18. H. Buchner et al., Inverse localization of electric dipole current sources in finite element models of the human head, *Electroencephalography and Clinical Neurophysiology*, vol. 102.4, pp. 267–278, 1997.

19. S. Baillet et al., Evaluation of inverse methods and head models for EEG source localization using a human skull phantom, *Physics in Medicine and Biology*, vol. 46.1, p. 77, 2001.

20. H. Jens et al., The influence of brain tissue anisotropy on human EEG and MEG, *Neuroimage*, vol. 15.1, pp. 159–166, 2002.

21. J. Haueisen, C. Ramon, P. Czapski, and M. Eiselt, On the influence of volume currents and extended sources on neuromagnetic fields: A simulation study, *Annals of Biomedical Engineering*, vol. 23, pp. 728–739, 1995.

22. G. Pruis, B. H. Gilding, and M. Peters, A comparison of different numerical methods for solving the forward problem in EEG and MEG, *Physiological Measurement*, vol. 14, p. A1, 1993.

23. G. Marin et al., Influence of skull anisotropy for the forward and inverse problem in EEG: Simulation studies using FEM on realistic head models, *Human Brain Mapping*, vol. 6.4, pp. 250–269, 1998.

24. T. J. R. Hughes, *The Finite Element Method: Linear Static and Dynamic Finite Element Analysis*, North Chelmsford, MA: Courier Corporation, 2012.

25. G. F. Carrier, K. Max, and C. E. Pearson, *Functions of a Complex Variable: Theory and Technique*, Philadelphia, PA: Society for Industrial and Applied Mathematics, 2005.

26. ANSYS. Available: http://www.ansys.com/. Accessed on January 16, 2017.

27. BESA. Available: http://www.besa.de/. Accessed on January 16, 2017.

28. SimBio Consortium, *SimBio: A generic environment for bio-numerical simulation*, 2000.

29. J. Haueisen, A. Büttner, H. Nowak, H. Brauer, and C. Weiller, The influence of conductivity changes in boundary element compartments on the forward and inverse problem in electroencephalography and magneto-encephalography [Der Einfluß der Änderung der Schalenleitfähigkeit bei Randelementemodellen auf die Vorwärtsrechnung und das inverse Problem in Elektroenzephalographie und Magnetoenzephalographie], *Biomedizinische Technik/Biomedical Engineering*, vol. 44, pp. 150–157, 1999.

30. D. B. Geselowitz, On bioelectric potentials in an inhomogeneous volume conductor, *Biophysical Journal*, vol. 7, p. 1, 1967.

31. G. Adde, M. Clerc, O. Faugeras, R. Keriven, J. Kybic, and T. Papadopoulo, Symmetric BEM formulation for the M/EEG forward problem, *Lecture Notes in Computer Science*, pp. 524–535, 2003.

32. A. Gramfort, T. Papadopoulo, E. Olivi, and M. Clerc, OpenMEEG: Opensource software for quasistatic bioelectromagnetics, *Biomedical Engineering Online*, vol. 9, p. 45, 2010.

33. Statistical Parametric Mapping. Available: http://www.fil.ion.ucl.ac.uk/spm/. Accessed on January 16, 2017.

34. Brainstorm. Available: http://neuroimage.usc.edu/brainstorm/. Accessed on January 16, 2017.

35. H. Hallez, B. Vanrumste, P. Van Hese, S. Delputte, and I. Lemahieu, Dipole estimation errors due to differences in modeling anisotropic conductivities in realistic head models for EEG source analysis, *Physics in Medicine and Biology*, vol. 53, p. 1877, 2008.

36. L. M. Harrison, W. Penny, J. Ashburner, N. Trujillo-Barreto, and K. Friston, Diffusion-based spatial priors for imaging, *NeuroImage*, vol. 38, pp. 677–695, 2007.

chapter five

EEG inverse problem I
Classical techniques

Introduction

This chapter discusses the classical techniques that are used to solve the EEG inverse problem. Before going into the discussion, let us revise the various elements for the inverse problem that were mentioned in previous chapters. The nature of the EEG source localization system matrix is rank deficient, which indicates that it has an infinite number of solutions. According to Golub [1], suppose $A \in \Re^{m \times n}$ and rank $(A) = r < n$, then this rank-deficient matrix is solved using various techniques. This rank deficiency is also termed *ill-posedness*, as defined by Hadamard in the last century. It was explained by Hadamard that if the solution is not unique or discontinues in nature, then it is termed as ill-posed in nature. These systems are termed *underdetermined systems*, as the number of known quantities (i.e., EEG sensors) is in the range of 128, 256, or 512; however, the number of unknowns (i.e., active sources throughout the cortex) is on the order of 10,000, which leads to severity of the underdetermined problem, and hence, regularization is needed to restrict the range of allowable solutions. According to the standard definition provided in Golub and Van Loan [2], if x is a minimizer and $z \in \text{null}(A)$, then $x + z$ is also a minimizer. Hence, the set of all minimizers is given as:

$$X = \{x \in \Re^n : \| Ax - b \|_2 = \min\} \tag{5.1}$$

If X is convex, and assuming $x_1, x_2 \in X$ and $\lambda \in [0,1]$, then

$$\| A[\lambda x_1 + (1-\lambda)x_2] - b \|_2 \leq \lambda \| Ax_1 - b \|_2 + (1-\lambda) \| Ax_2 - b \|_2 \\ = \min \| Ax - b \|_2 \tag{5.2}$$

Hence, it is obvious that X has a unique element having a minimum two norm. This solution is the minimum norm solution. This gives rise to the development of a solution for inverse problems for an ill-posed EEG system, as we see in later sections. Hence, a class of methods is based on minimum norm estimation (MNE) and its variant forms for inverse problem solution. Now we shall move to the discussion pertaining to inverse problems that are formulated generally as follows:

$$\int_{\Omega} \text{Input} \times \text{System } d\Omega = \text{Output} \qquad (5.3)$$

This formulation suggests computing the output given the input and mathematical derivations for the system. In other words, the inversion techniques target finding the input or the system that is responsible for the production of output. For example, in our case, we have output signals measured through a set of sensors—that is, potentials—and the input signals are the primary currents (or associated volume currents), which cause some noise (error in measurements) in the recorded EEG data.

According to Lawson and Hanson [3], the classical example of a linear ill-posed problem is the Fredholm integral equation, which is given as follows:

$$\int_{0}^{1} K(s,t)f(t)dt = g(s), \quad 0 \le s \le 1 \qquad (5.4)$$

where kernel K and g are known quantities, whereas f is unknown for which the solutions are sorted out using minimization algorithms as stated earlier. For most practical problems, including EEG inverse problems, the kernel K is defined by the underlying mathematical model; however, g has got the measured quantities within it. This implies that g has known values with certain accuracy along with certain noise levels. In comparison with the EEG inverse problem, here K is the leadfield matrix defined by head modeling schemes and system (head) dynamics, such as tissue conductivities, permeability values. However, the g is voltage measured through an EEG sensor cap with a certain sort of noise imparted due to various impairments. Another basic definition for the ill-posed systems is through singular value decomposition (SVD), which says that if the SVD is taken for a system matrix having ill-posedness, then the singular values are decreasing in nature. In other words, the discrete ill-posed problem should satisfy the following conditions [4,5]:

1. The singular values of system matrix (A) should be decaying gradually to zero.
2. The ratio between the largest and the smallest nonzero singular values is large.

When solving the underdetermined or ill-posed problems, it is necessary to incorporate further information about the desired solution to stabilize the problem and estimate useful and stable solutions. This is achieved by imparting regularization into the solution. Among the various regularization methods, the most cited and common is the Tikhonov regularization method [6,7]. The main concept of Tikhonov regularization is to define a regularized solution x_λ as the minimizer of the weighted combination of the residual norm and the side constraint, such that

$$x_\lambda = \arg\min\left\{\|Ax - b\|_2^2 + \lambda^2 \|L(x - x^*)\|_2^2\right\} \tag{5.5}$$

where λ is the regularization parameter, which controls the weight applied for minimization of the side constraint relative to minimization of the residual norm. It is evident that larger values of regularization parameter provide small-solution seminorm at the cost of a large residual norm, and vice versa. Thus, following this strategy for the solution of ill-posed problems, the EEG inverse problem is solved using least squares, minimum norms, weighted minimum norm (WMN), and SVD-based subspace methods. Before providing further explanation of such methods, we here first define the generalized linear model (GLM) for EEG measurements. The GLM for EEG is written as follows:

$$\mathbf{Y} = \mathbf{L}\mathbf{J} + \in \tag{5.6}$$

where $\mathbf{Y} \in \Re^{N_c \times N_n}$ is the dataset acquired by N_c sensors, and the number of time samples is N_n. The current density $\mathbf{J} \in \Re^{N_d \times N_n}$ is responsible for the propagation of the energy of N_d current dipoles distributed through the cortical surface. However, the dataset ($\mathbf{Y}$) and the sources ($\mathbf{J}$) are related through gain matrix $\mathbf{L}$, which is also termed the *leadfield matrix*. Thus, the assumptions of zero mean Gaussian noise $\in$ is considered with covariance $\mathbf{Q}_\in$. The source localization considers the activation images of the brain by searching the magnitudes of current dipoles ($\mathbf{J}$), which fit the data more closely. Because the number of dipoles is greater than the number of sensors—that is, $N_d \gg N_c$—the leadfield matrix becomes noninvertible, defining the ill-posedness of the system. This leadfield matrix is a dense mesh of elemental cortical dipoles. As the sensor array for EEG data capture includes 100–300 sensors, the system is under highly constrained conditions.

Hence, by comparing GLM with the linear system equation—that is, $Ax = b$—it can be deduced that $\mathbf{Y}$, which is the measured potential, is analogous to b, whereas the leadfield matrix $\mathbf{L}$ is related to A and the unknown quantities, which are supposed to be evaluated through different methods, are x in the linear system and current density ($\mathbf{J}$) in terms of the EEG inverse problem, respectively.

Now we move forward with our discussion of classical inversion techniques. After this, we discuss the founding method, which is the MNE, and then move on to low-resolution brain electromagnetic tomography (LORETA), standardized LORETA (sLORETA), exact LORETA (eLORETA), and finally end our chapter with the beamforming method. The discussion for each method is supported by proper mathematical formulations and citations.

5.1 Minimum norm estimation

This solution to the EEG inverse problem was proposed by M. S. Hämäläinen and R. J. Ilmoniemi in 1984. This is the most generalized approach to estimate 3D source distribution in the absence of any *a priori* information. The only assumption it takes into account for a solution is that the current distribution should have a minimum overall intensity (smallest L2-norm) [8]. The idea to use minimum norm estimates was first provided by M. S. Hämäläinen et al. in "Interpreting Magnetic Fields of the Brain: Minimum Norm Estimates" [9]. According to Hämäläinen et al. [10], the MNE provides best estimates when less *a priori* information about the source distribution is available. In the mentioned article [9,10], the magnetic fields were discussed at length; however, the same analogy can also be derived for electric fields. The analysis starts by explaining the leadfield as the linear combination of the magnetometer leadfield that provides the estimates for primary current source distribution in the brain. Thus, the primary current density is expressed as:

$$\mathbf{J}^{\mathbf{P}}(r) = \mathbf{J}_{tot}(r) - \sigma(r)\mathbf{E}(r) \tag{5.7}$$

where r is position vector, $\mathbf{J}_{tot}$ is the total current density, σ is the conductivity, and $\mathbf{E}$ is the electric field. The primary current is due to the change in other forms of energy into electrical. However, the second term of the equation $\sigma\mathbf{E}$ is the volume current within the conductor. The output of magnetometer $\mathbf{B}_i$ is linearly related to the primary current distribution such that

$$\mathbf{B}_i = \int \mathbf{L}_i(r) \cdot \mathbf{J}^{\mathbf{P}}(r) dv \tag{5.8}$$

In Equation 5.8, $\mathbf{L}_i(r)$ is the leadfield, which defines the sensitivity distributed for the ith magnetometer. The leadfield calculation is dependent on the conductivity σ, and thus can be estimated, provided the solution of $\mathbf{B}$ is calculated.

Continuing the discussion for the estimation of source distribution, here a subspace F' is defined, which is derived from function space F such that all primary currents are elements of it. Mathematically,

$$\mathbf{J}_1 \in F, \mathbf{J}_2 \in F \tag{5.9}$$

The inner product as provided in Equation 5.8 can provide information only about primary currents present in subspace F' of the current
space F. This subspace F' is spanned by leadfields such that

$$F' = \mathrm{span}(L_1, L_2, ..., L_M) \tag{5.10}$$

Hence, MNE is used to find the estimated current density $\mathbf{J}'$ for $\mathbf{J}^P$,
which is restricted to subspace F'. Hence, $\mathbf{J}'$ will be a linear combination of
the leadfields such that

$$\mathbf{J}' = w^T \mathbf{L} \tag{5.11}$$

where w is weighting scalars that are evaluated through measurements.
From the equation, it is evident that for the estimation of $\mathbf{J}'$, a set of linear
equations is obtained such that

$$b = \Gamma w \tag{5.12}$$

where $b = [B_1, ..., B_M]^T$, $w = [w_1, ..., w_M]^T$, and $\Gamma \in \Re^{M \times M}$ has inner products
of leadfields such that $\Gamma_{ij} = \langle L_i, L_j \rangle$.

The eigenvalue decomposition (EVD) of Γ reveals that it has very
small eigenvalues, which contribute the largest error for computation of
w. To remove this instability in solution, the regularization is imparted
in solution, which states that the eigenleads having smaller eigenvalues
are ignored as they have less signal-to-noise values. Hence, with some
manipulation related to eigenvalue decomposition, the estimated solution
for MNE with regularization is given as:

$$\tilde{\mathbf{J}}' = (\tilde{\Gamma}^{-1}b)^T \mathbf{L} \tag{5.13}$$

In this book, the MNE has been applied for various conditions
such as with and without noise and for evoked responses. Our earlier
discussion revealed that MNE can be improved by imparting some *a
priori* knowledge into the solution. It was also suggested to increase the
number of magnetometers (which are sensors in the case of EEG-based
source localization) to improve the overall quality of the solution. In
addition, the noise effect was suggested to be reduced by imparting
spatial smoothening of the outputs. Although MNE offers good results
in terms of resolution and current estimation, it is unable to address

the issue of deep source localization in the outermost cortex. This is because M/magnetoencephalography is a harmonic function, that is, $\nabla^2 J = 0$; and the harmonic functions attain maximum values at the boundaries of their domain, which in this case is the outermost cortex [11]. In addition, upon comparison with newer techniques such as LORETA and WMN, the minimum norm solution has more localization errors as well as has the disadvantage of being incapable of localizing nonboundary sources.

5.2 Low-resolution brain electromagnetic tomography

This technique was introduced by R. D. Pascual-Marqui [12–14]. LORETA calculates current distribution throughout the brain volume. This technique was proposed to localize the 3D solutions properly as compared with previous minimum norm approaches for which there was no prior knowledge. However, LORETA imposes a spatial smoothness constraint upon solution. This spatial smoothness constraint is expressed using the 3D discretized Laplacian matrix. Thus, LORETA produces a discrete, 3D-distributed, linear inverse solution with improved time resolution but with low spatial resolution. For the low-noise instantaneous measurements, the discrete solution for LORETA is derived as:

$$\min_{J} \left\| \mathbf{B}W\mathbf{J} \right\|^2 \quad \text{under constraint} \quad \mathbf{Y} = \mathbf{L}\mathbf{J} \tag{5.14}$$

where weighted matrix W is defined as $W = \Omega \otimes I$, $I \in \Re^{3 \times 3}$ is identity matrix and $\otimes$ defines the Kronecker product, Ω is a diagonal matrix for which the diagonal element is defined by

$$\Omega_{ii} = \sqrt{\sum_{\alpha=1}^{N} L_{\alpha i}^T L_{\alpha i}} \tag{5.15}$$

The discrete Laplace operator $\mathbf{B}$ is introduced to emphasize relationships between current densities, and thus, the spatial resolution is not taken into consideration, which results in blurred localization images. For a regular cubic grid within the brain volume, if the distance between neighboring grids is assumed to be d, then this operator is defined as:

$$\mathbf{B} = \frac{6}{d^2}(A - I_{3M}) \tag{5.16}$$

Here, $A = A_0 \otimes I_3$ with $A_0 = 0.5 \times \{I_M + [diag(A_1 1_M)]^{-1}\} A_1$. M defines the 3D locations for the current densities. As per definition, the element in A_1 is defined as if the grid which serial number is α neighbor on the grid which serial number is β, then its value is 1/6, else is zero. Hence

$$A_1 = \begin{bmatrix} a_{11} & \cdots & a_{1M} \\ \vdots & & \vdots \\ a_{M1} & \cdots & a_{MM} \end{bmatrix}$$

where

$$a_{\alpha\beta} = \begin{cases} \dfrac{1}{6} & \text{if } \|r_\alpha - r_\beta\| = d \\ 0 & \end{cases} \quad \forall \alpha, \beta = 1, \ldots, M \tag{5.17}$$

Hence, the solution is provided as:

$$\mathbf{J}' = T\mathbf{Y}, \text{ with } \quad T = (WB^T BW)^{-1} L^T \{L(WB^T BW)^{-1} L^T\}^+ \tag{5.18}$$

where A^+ denotes the Moore–Penrose pseudoinverse of matrix A.

The inverse solution is valid for both electric and magnetic data taken separately or simultaneously. It should be noted that this equation for inverse solution holds equal for all head modeling schemes provided that the leadfield is calculated properly. The usage of constraints can impart anatomical and physiological priors to make the solution more stable with higher resolution. Hence, in this way, LORETA finds the active sources by minimizing the Laplacian of weighted sources and selecting the solution with smooth spatial distribution. Because of such smoothness for the solution, LORETA provides the solution with low spatial resolution—that is, blurred solution (over smoothed), which includes two hemispheres or different lobes.

This algorithm is tested extensively on various datasets and a software package with the name LORETA has been developed at The Key Institute for Brain-Mind Research, University Hospital of Psychiatry, University of Zurich, Switzerland [15]. In addition, a variety of other methods have been developed by R. D. Pascual-Marqui, such as sLORETA and eLORETA, which have the same basic definition as LORETA. The main advantage with LORETA is that it can provide the basic solution for localization with easy-to-follow procedures as it is based on weighted Laplacian only. In addition, unlike the MNE solution, LORETA can localize the boundary and deep sources. However, the solution provided has low spatial resolution, which is an undesirable feature when we are dealing with pattern-recognition applications of brain source localization.

5.3 Standardized LORETA

This is the first variation on LORETA, which was proposed by the inventor of LORETA (R. D. Pascual-Marqui) [16]. This method is based on standardization of current density that is supposed to be estimated for source localization. This method has some resemblance to the method proposed by Dale [17] where the localization inference was based on standardization of the current density. Hence, the current density estimate was carried out using the MNE approach, and it was further standardized using its expected standard deviation, which is hypothesized to have originated exclusively by noise in measurements. The approach adopted by Dale produced systematic nonzero localization errors even in the presence of minor noise levels. However, sLORETA is similar to the method proposed by Dale et al. as it provides current density estimates using the MNE solution, and the localization inference is based on standardized values of the current density estimates. The way sLORETA adopts the standardization for current density is different; thus, the result is lesser localization error as compared with the Dale method.

According to the standard definition for minimum norm estimates, the solution provided by sLORETA is harmonic in nature, which implies that Laplacian of current density is zero (i.e., $\nabla^2 J[r] \equiv 0$), where r denotes the volume coordinates in the brain. Thus, it produces a smooth solution as mentioned in the "Minimum Norm Estimation" section. As a result, the deep point sources in the brain are not properly localized and localization error is relatively high. This problem is solved through standardization of the minimum norm solution, and thus, by making localization inference on such a standardization.

Using the same definitions for the variables as mentioned in previous methods, the functional interest is given as follows:

$$F = \left\| \mathbf{Y} - \mathbf{L}\mathbf{J} - c_1 \right\|^2 + \alpha \left\| \mathbf{J} \right\|^2 \tag{5.19}$$

Here, $\alpha \geq 0$ is the regularization parameter. This functional is supposed to be minimized with respected to $\mathbf{J}$ and c for a given leadfield ($\mathbf{L}$), voltage measurements ($\mathbf{Y}$), and regularization parameter (α). The explicit solution for such minimization will be

$$\mathbf{J}' = T\mathbf{Y}, \text{ with } \quad T = L^T H [HLL^T H + \alpha H]^+ \tag{5.20}$$

where $H \in \Re^{N_E \times N_E}$ is the centering matrix and is defined as:

$$\mathbf{H} = I - \frac{11^T}{1^T 1} \tag{5.21}$$

However, $I \in \Re^{N_E \times N_E}$ is the identity matrix, and $1 \in \Re^{N_E \times 1}$ is a vector of ones.

Using the average reference for the measurements, one will have the functional as:

$$F = \|\mathbf{Y} - \mathbf{LJ}\|^2 + \alpha \|\mathbf{J}\|^2 \tag{5.22}$$

with solution as:

$$\mathbf{J}' = T\mathbf{Y}, \text{ with } \quad T = L^T [\mathbf{LL}^T + \alpha \mathbf{H}]^+ \tag{5.23}$$

For the standardization of the solution, that is, $\mathbf{J}'$, we need to calculate its variance. According to the Bayesian formulation, the actual source variance (prior), $S_J \in \Re^{(3Nv) \times (3Nv)}$, is equal to the identity matrix:

$$S_J = I, I \in \Re^{(3Nv) \times (3Nv)} \tag{5.24}$$

Here, N_v is the number of voxels in the brain.

Besides the potential measurements, which are due to noisy measurements, will give variance as:

$$S_Y^{noise} = \alpha \mathbf{H} \tag{5.25}$$

Hence, making use of the aforementioned equations and considering the actual source activity and measurement noise independently, the electric potential variance is defined as:

$$S_Y = \mathbf{L}S_J L^T + S_v^{noise} = \mathbf{LL}^T + \alpha \mathbf{H} \tag{5.26}$$

Thus, the variance of estimated current density will be

$$S_{J'} = TS_Y T^T = T(\mathbf{LL}^T + \alpha \mathbf{H})T^T = L^T [\mathbf{LL}^T + \alpha \mathbf{H}]^+ L \tag{5.27}$$

Hence, the estimated current density for sLORETA is given by

$$\mathbf{J}' = TLJ = L^T (\mathbf{LL}^T + \alpha \mathbf{H})^+ LJ = S_{J'} \mathbf{J} \tag{5.28}$$

In this way, the sources are estimated in the presence and absence of noise. It has been claimed [16], which had different setups for EEG data, that sLORETA estimates the current sources without any localization error—that is, it provides the exact solution with zero localization error. Hence, it was termed as the best first-order source localization technique. The simulations with and without noise were performed to prove the claim. Hence, it was suggested that sLORETA has better quality as compared with minimum norm and Dale methods. This method has been used extensively for source localization with various head modeling schemes. Different neuroimaging software implement this technique, including NetStation, LORETA, and so on. Many publications followed the sLORETA technique to localize the sources. The disadvantage with sLORETA is its low resolution, which is due to regularization in solution for stability. In addition, sLORETA fails to localize the multiple sources when the point-spread functions of sources overlap.

5.4 Exact LORETA

There have been many useful attempts to minimize the localization error by choosing the weight matrix in a more adequate way. However, there exists one methodology to give more importance to the deeper sources with reduced localization error, which is termed as eLORETA. This method was developed and officially recorded as a working project at the University of Zurich in March 2005 [18]. The idea is mainly based on the LORETA method except for the difference in implementation as we see now. Consider the generalized WMN solution as:

$$\mathbf{J}' = T\mathbf{Y} \text{ with } \quad T = W^{-1}L^T[LW^{-1}L^T + \alpha\mathbf{H}]^+ \tag{5.29}$$

where $W \in \Re^{(3N_V) \times (3N_V)}$ is the symmetric weight matrix and $\alpha \geq 0$ is the regularization parameter as defined above.

A particular case is considered for the weight matrix such that only it has nonzero diagonal entities, so $W_i \in \Re^{3 \times 3}$.

Hence, for the regularization parameter, that is, $\alpha = 0$, the estimated current density at ith voxel is given by

$$j_i' = W_i^{-1}L_i^T[LW^{-1}L^T + \alpha\mathbf{H}]^+ \tag{5.30}$$

Hence, with the solution equation as provided above, the weights are defined as follows:

$$W_i = [L_i^T(LW^{-1}L^T + \alpha\mathbf{H})^+ L_i]^{1/2} \tag{5.31}$$

Furthermore, the method that satisfies the weights provided by the equation is termed as eLORETA, which according to their inventors is the exact method with zero localization error. In addition, it should be noted that eLORETA is standardized in nature like sLORETA, which indicates that its theoretical expected variance is unity. eLORETA is unbiased in the presence of measurement and structural biological noise, which is expressed as:

$$\text{cov}(\epsilon_j) = \sigma_j W^{-1} \tag{5.32}$$

Thus, it was concluded in the aforementioned article using different formulations that eLORETA is the perfect solution to the M/EEG inverse problem, which has zero localization error in the presence of measurement and structural biological noise. In addition, it was mentioned that eLORETA depends on data, which show that it is an adaptive technique.

The simulations for the validation of this method were carried out under free academic eLORETA-KEY software. The results indicated that eLORETA is a reliable localizing method with no localization bias, providing zero error localization in the case of nonideal conditions—that is, the presence of noise. There are several studies providing comparisons between sLORETA and eLORETA, which show that eLORETA provides better results than sLORETA [19].

5.5 *Focal underdetermined system solution*

The focal underdetermined system solution (FOCUSS) is a high-resolution, nonparametric recursive algorithm that is used to reconstruct the brain sources in a recursive manner. This algorithm has been defined and developed [20], and states that it is based on the formulation of WMN in a recursive order, resulting in high resolution but at the cost of high computational complexity. FOCUSS starts to localize the sources with a distributed estimate such as a minimum norm at the initial stage, after which it recursively enhances the values of some of the initial solution elements, while decreasing the rest of the elements until they become zero. This procedure is carried out in a repetitive manner. This process generates only a small number of winning nonzero elements, which provides a solution with high resolution for EEG data.

The numerical solution for FOCUSS follows WMN. Although it has a pseudoinverse solution like WMN, it includes an extra constraint or weight on the solution. The computational equation is written as:

$$\mathbf{I} = W(GW)^{+}\mathbf{B} = WW^{T}G^{T}(GWW^{T}G^{T})^{-1}\mathbf{B} \tag{5.33}$$

Here, $\mathbf{I} \in \mathfrak{R}^{n \times 1}$ represents a vector having the unknown dipole amplitude as its entry. The inverse modeling can compute this unknown. However, $G \in \mathfrak{R}^{m \times n}$ is the basis matrix whose known elements are explicitly specified in the model. The elements of G are termed *spatial weighting elements* as each of them specifies the contribution to a given magnetic measurement from one dipole component of unit strength. Because intracranial magnetic permeability is considered constant, the elements of G are pure geometric factors. As G is constant and can be precalculated for a particular head and sensor geometry, it is feasible to input very complex and detailed forward models incorporating realistic geometry and major compartments within the head volume conductance. This is analogous to the leadfield matrix in previous discussions where different head models were used for a particular conductivity to compute the forward model. In addition, $\mathbf{B} \in \mathfrak{R}^{m \times 1}$ represents the vector having m radial magnetic field measurements at the m sensors. The mathematical relationship between these variables is as follows:

$$\begin{bmatrix} \mathbf{B} \\ \vdots \\ \mathbf{B}_m \end{bmatrix} = \begin{bmatrix} G_{1,x}, & G_{1,y}, & G_{1,z}, & G_{2,x} & \cdots & G_{n/3,z}, \end{bmatrix} \cdot \begin{bmatrix} I_1 \\ I_2 \\ \vdots \\ I_n \end{bmatrix} \qquad (5.34)$$

In Equation 5.33, $W \in \mathfrak{R}^{n \times n}$ is a weight matrix, which provides constraint on the solution. This so-called regularization enhances some of the elements in $\mathbf{I}$. The recursive steps of FOCUSS change weight matrix W. For the enhancement of some of the already prominent elements, W is constructed directly from the amplitudes of the elements of $\mathbf{I}$ of the preceding steps. Such W matrices are dimensionless scaling factors, which can be varied to produce any number of recursive schemes. Furthermore, W can be constructed by taking its diagonal elements to be the previous iterative step solution, such that

$$W_k = \begin{bmatrix} \mathbf{I}_{1_{K-1}} & & 0 \\ & \ddots & \\ 0 & & \mathbf{I}_{n_{K-1}} \end{bmatrix} \qquad (5.35)$$

where $\mathbf{I}_{i_{K-1}}$ represents the ith element of the vector $\mathbf{I}$ at the $(k-1)$ iteration, and k is the index of the iteration step.

Hence, the FOCUSS algorithm follows the following steps:

1. Construction of weight matrix W from the corresponding elements of the previous step from the equations defined above. Thus, the constraints are included in the W.

2. Computation of the WMN solution with Equation 5.33 using *W* as calculated in Step 1.

These steps are repeated until solution **I** no longer changes, which shows the convergence of the algorithm.

Various publications have followed the FOCUSS with and without noise levels [21–30]. The procedure is straightforward with an assumption of no *a priori* information for source localization. However, due to the large iteration and recursive nature, the error is imparted, which results in unstable solution. To overcome this limitation, regularization is employed for FOCUSS. It has been reported in the literature that FOCUSS has considerable utility for the class of inverse problems in which localized sources are expected. Hence, it can serve as a valuable tool either on its own or as a part of a larger imaging strategy that combines a number of techniques.

Summary

This chapter discussed classical techniques that are used for brain source localization. Hence, the discussion was started by providing mathematical background related to the inverse problem in general and to these classical techniques in particular. Thus, MNE was defined and explained with the aid of derivations to provide a clearer picture of this initial method. The discussion moved forward to define LORETA, which is an advanced version of MNE and works on second-order Laplacian to provide a distributed solution with low spatial resolution. After this, the standardized version of LORETA, sLORETA was explained in detail. The latest version of the LORETA family eLORETA was explained after sLORETA. Thus, the chapter ended by discussing the FOCUSS method, which is considered to belong to the same classical group as it employs WMN for the source estimation. It is concluded that MNE is the simplest method with the disadvantage of inefficiency for deep and boundary sources. However, the LORETA family has low resolution, which is not a desirable feature when used for pattern recognition. The FOCUSS is simple in implementation but has a recursive nature, which makes it computationally complex.

References

1. G. Golub, Numerical methods for solving linear least squares problems, *Numerische Mathematik*, vol. 7, pp. 206–216, 1965.
2. G. H. Golub and C. F. Van Loan, *Matrix Computations*, Vol. 3, Baltimore, MD: JHU Press, 2012.
3. C. L. Lawson and R. J. Hanson, *Solving Least Squares Problems*, Vol. 161, Philadelphia, PA: Society for Industrial and Applied Mathematics, pp. 161, 1974.

4. P. C. Hansen, Regularization tools: A Matlab package for analysis and solution of discrete ill-posed problems, *Numerical Algorithms*, vol. 6(1), pp. 1–35, 1994.

5. P. C. Hansen and D. P. O'Leary, The use of the L-curve in the regularization of discrete ill-posed problems, *SIAM Journal on Scientific Computing*, vol. 14(6), pp. 1487–1503, 1993.

6. C. W. Groetsch, *The Theory of Tikhonov Regularization for Fredholm Equations*, Boston, MA: Boston Pitman Publication, 1984.

7. G. H. Golub, P. C. Hansen, and D. P. O'Leary, Tikhonov regularization and total least squares, *SIAM Journal on Matrix Analysis and Applications*, vol. 21(1), pp. 185–194, 1999.

8. P. C. Hansen, Rank-deficient and discrete ill-posed problems: Numerical aspects of linear inversion. *Society for Industrial and Applied Mathematics*, vol. 3(4), pp. 253–315, 1998.

9. M. S. Hämäläinen and R. J. Ilmoniemi, Interpreting magnetic fields of the brain: Minimum norm estimates. *Medical and Biological Engineering and Computing*, vol. 32(1), pp. 35–42, 1994.

10. M. Hämäläinen, R. Hari, R. J. Ilmoniemi, J. Knuutila, and O. V. Lounasmaa, Magnetoencephalography—Theory, instrumentation, and applications to noninvasive studies of the working human brain, *Reviews of Modern Physics*, vol. 65, p. 413, 1993.

11. M. Jatoi, N. Kamel, A. Malik, and I. Faye, EEG based brain source localization comparison of sLORETA and eLORETA, *Australasian Physical & Engineering Sciences in Medicine*, vol. 37, pp. 713–721, 2014.

12. R. D. Pascual-Marqui, C. M. Michel, and D. Lehmann, Low resolution electromagnetic tomography: A new method for localizing electrical activity in the brain. *International Journal of Psychophysiology*, vol. 18(1), pp. 49–65, 1994.

13. R. D. Pascual-Marqui, Review of methods for solving the EEG inverse problem, *International Journal of Bioelectromagnetism*, vol. 1, pp. 75–86, 1999.

14. R. D. Pascual-Marqui et al., Low resolution brain electromagnetic tomography (LORETA) functional imaging in acute, neuroleptic-naive, first-episode, productive schizophrenia, *Psychiatry Research: Neuroimaging*, vol. 90, pp. 169–179, 1999.

15. sLORETA. Available: http://www.unizh.ch/keyinst/NewLORETA/sLORETA-Math01.pdf. Accessed on January 16, 2017.

16. R. D. Pascual-Marqui, Standardized low-resolution brain electromagnetic tomography (sLORETA): Technical details, *Methods and Findings in Experimental and Clinical Pharmacology*, vol. 24, pp. 5–12, 2002.

17. A. M. Dale et al., Dynamic statistical parametric mapping: Combining fMRI and MEG for high-resolution imaging of cortical activity, *Neuron*, vol. 26, pp. 55–67, 2000.

18. R. D. Pascual-Marqui, Discrete, 3D distributed, linear imaging methods of electric neuronal activity. Part 1: exact, zero error localization. arXiv preprint arXiv:0710.3341, 2007.

19. M. Jatoi, N. Kamel, A. Malik, and I. Faye, EEG based brain source localization comparison of sLORETA and eLORETA, *Australasian Physical & Engineering Sciences in Medicine*, vol. 37, pp. 713–721, 2014.

20. I. F. Gorodnitsky, J. S. George, and B. D. Rao, Neuromagnetic source imaging with FOCUSS: A recursive weighted minimum norm algorithm, *Electroencephalography and Clinical Neurophysiology*, vol. 95, pp. 231–251, 1995.

21. K. Rafik, B. H. Ahmed, F. Imed, and T.-A. Abdelmalik, Recursive sLORETA-FOCUSS algorithm for EEG dipoles localization, in *First International Workshops on Image Processing Theory, Tools and Applications, 2008 (IPTA 2008)* Capri Island, Italy: IEEE, 2008, pp. 1–5.

22. L. He Sheng, Y. Fusheng, G. Xiaorong, and G. Shangkai, Shrinking LORETA-FOCUSS: A recursive approach to estimating high spatial resolution electrical activity in the brain, in *First International IEEE EMBS Conference on Neural Engineering, 2003*, New York: IEEE, 2003, pp. 545–548.

23. L. Hesheng, P. H. Schimpf, D. Guoya, G. Xiaorong, Y. Fusheng, and G. Shangkai, Standardized shrinking LORETA-FOCUSS (SSLOFO): A new algorithm for spatio-temporal EEG source reconstruction, *IEEE Transactions on Biomedical Engineering*, vol. 52, pp. 1681–1691, 2005.

24. I. F. Gorodnitsky and B. D. Rao, Sparse signal reconstruction from limited data using FOCUSS: A re-weighted minimum norm algorithm. *IEEE Transactions on Signal Processing*, vol. 45(3), pp. 600–616, 1997.

25. H. Jung et al., k-t FOCUSS: A general compressed sensing framework for high resolution dynamic MRI, *Magnetic Resonance in Medicine*, vol. 61(1), pp. 103–116, 2009.

26. S. M. Bowyer et al., MEG localization of language-specific cortex utilizing MR-FOCUSS. *Neurology*, vol. 62(12), pp. 2247–2255, 2004.

27. J. E. Moran, S. M. Bowyer, and N. Tepley, Multi-resolution FOCUSS: A source imaging technique applied to MEG data, *Brain Topography*, vol. 18(1), pp. 1–17, 2005.

28. J. Han and K. S. Park, Regularized FOCUSS algorithm for EEG/MEG source imaging, in *26th Annual International Conference of the IEEE Engineering in Medicine and Biology Society, 2004 (IEMBS'04)*, Vol. 1, San Francisco, US: IEEE, 2004.

29. I. F. Gorodnitsky, and B. D. Rao, Sparse signal reconstruction from limited data using FOCUSS: A re-weighted minimum norm algorithm, *IEEE Transactions on signal processing*, vol. 45(3), pp. 600–616, 1997.

30. A. Majumdar, FOCUSS based Schatten-p norm minimization for real-time reconstruction of dynamic contrast enhanced MRI. *IEEE Signal Processing Letters*, vol. 19(5), pp. 315–318, 2012.

chapter six

EEG inverse problem II
Hybrid techniques

Introduction

This chapter discusses some of the hybrid techniques based on a combination of previously developed algorithms. First, the hybrid weighted minimum norm (WMN) is discussed with adequate mathematical background. A hybrid algorithm of WMN and low-resolution brain electromagnetic tomography (LORETA) is then discussed, followed by the recursive technique based on hybridization of standardized LORETA (sLORETA) and focal underdetermined system solution (FOCUSS) techniques. Furthermore, another hybrid technique, which is based on a combination of LORETA and FOCUSS, termed shrinking LORETA-FOCUSS is elaborated. Finally, the discussion is completed with an explanation of standardized shrinking LORETA-FOCUSS, which is a combination of LORETA and FOCUSS with some exceptions, which are discussed later.

6.1 Hybrid WMN

This algorithm was proposed and explained in Song, Wu, and Zhuang [1]. It takes advantage of the low resolution provided by LORETA, which emphasizes localization only, and high resolution provided by FOCUSS, which emphasizes separability. The basic framework is based on the WMN strategy, where the construction of the weight matrix is achieved by taking reference from existing smoothing operator. Hence, both LORETA and FOCUSS are used to localize brain activity in a way that LORETA is used to get initial source reconstruction. Because LORETA has low spatial resolution, the resultant reconstruction is a blur in nature.

The discrete result for the inverse problem in this algorithm with the initialization of LORETA gives the estimated current density as:

$$\hat{\mathbf{J}} = (WB^T\mathbf{B}W)^{-1}L^T \left(\mathbf{L}(WB^T\mathbf{B}W)^{-1}L^T\right)^+ \mathbf{Y} \tag{6.1}$$

However, by including the regularization term (λ) in this equation for current density to introduce stability, we can rewrite Equation 6.1 as:

$$\hat{\mathbf{J}} = (WB^T\mathbf{B}W)^{-1}G^T(G(WB^T\mathbf{B}W)^{-1}G^T + \lambda\mathbf{H})^+\mathbf{Y} \tag{6.2}$$

where all the parameters are as defined earlier.

Hence, using the iterative weighted method, under the constraint condition of the EEG forward equation $\mathbf{Y} = \mathbf{LJ}$ will result in intensification of some of the grid's energy in the solution space. The concrete way is taking the $\mathbf{J}_{k-1}$ step's solution as prior information to construct the k-step's weighted matrix W_k:

$$W_k = diag(\mathbf{J}_{k-1}) \tag{6.3}$$

Thus, using the WMN method, the kth iterative solution is

$$\mathbf{J}_k = W_k(\mathbf{L}W_k)^+\mathbf{Y} \tag{6.4}$$

Hence, this iterative procedure is repeated until convergence. The details of convergence are provided in Song, Wu, and Zhuang [1]. The simulations are performed by considering the four-shell homogeneous spherical head model, which represents the brain, cerebrospinal fluid, skull, and scalp, correspondingly, from inside to outside. The geometry parameter of the relative radius is (0.84, 0.8667, 0.9467, 1), and the physical parameter involved is the electrical conductivity whose corresponding values are (0.33, 1.0, 0.0042, 0.33) s/m. The solution space within the 3D brain volume is confined to a maximum radius of 0.84, with vertical coordinate value $z \geq -0.28$. There are 729 grid points within the solution space corresponding to a 3D regular cubic grid. Thus, with the two-dipole simulation, it was shown that with LORETA, a rough localization can be obtained. Furthermore, it was continued with the reweighted iterative method, which produced the exact solution with minimum error. Hence, the estimated dipole distribution is quite close to practical dipole distribution. Thus, the hybrid method provided the solution with less error and in an efficient way.

6.2 Weighted minimum norm–LORETA

This is a hybrid method used to localize the sources with possible minimum localization error. This hybrid technique makes use of the WMN technique, which was explained earlier, and the LORETA technique in a combinational way [2]. In this method, WMN is used to initialize the

LORETA such that the current density vector J_{WMN} is used as the initialization parameter for LORETA. Following this, initially, the current density is estimated with WMN using the following equation:

$$J_{WMN} = W^{-2}L^t(LW^{-2}L^t)^+Y \qquad (6.5)$$

where weight matrix W is defined as:

$$W_i = \left(\frac{1}{N_e}\right) \cdot \sqrt{\sum_{j=1}^{N_e} L_{ij}^2} \qquad (6.6)$$

where N_e is the number of electrodes.

After this stage, the weight matrix is constructed using the current density estimate as obtained in Equation 6.5. Hence, the weight matrix is given by

$$W_h = diag\left[J_{WMN}(i)\right] \qquad (6.7)$$

Hence, a new weight matrix is developed, which is derived from the WMN result, and thus is dependent on the WMN methodology. Therefore, the new weight matrix is given by

$$C_h = W_h B^t \mathbf{B} W_h \qquad (6.8)$$

Therefore, the final expression for the current density estimation using this hybrid method will be

$$J_{WMN-LORETA} = (C_h)^{-1}L^t[L(C_h)^{-1}L^t]^+Y \qquad (6.9)$$

In this way, WMN–LORETA provides a hybrid solution for source estimation.

This technique [2] was examined using 138 electrodes distributed on the scalp surface with 429 sources on the cerebral volume. The simulations for WMN, LORETA, and hybrid WMN–LORETA were performed to compare the results. It is observed that WMN presents a distribution of the current dipoles in-depth, but suppose that the neuronal activity is not regular. WMN–LORETA combines the advantages of the WMN and LORETA methods. The comparative study [2] was performed using the so-called resolution matrix (R). This matrix is used to analyze the methods to determine the qualities and limitations associated with each method.

Mathematically, it is defined as:

$$R = T\mathbf{L} \tag{6.10}$$

For an ideal situation, R is an identity matrix, which related the estimated current density and original current density as:

$$\mathbf{J}' = R\mathbf{J} \tag{6.11}$$

Hence, through the measurement of distance between resolution matrix (R) and identity matrix, we can measure the precision provided by certain methods. A comparison has been provided for WMN, LORETA, and WMN–LORETA in terms of resolution matrix and computational time, respectively [2]. It was observed that for WMN–LORETA, the resolution matrix (R) is close to the identity matrix, compared with other techniques. This shows better localization ability for the discussed method. However, the computational time taken by WMN–LORETA is just 2 seconds more than LORETA, which is a minor difference for such better precision. Hence, it was concluded that, for this research work, the developed WMN–LORETA method is efficient for the localization of sources having a highly focused activity, such as the somatosensory-evoked potentials and the analysis of epileptic brain activity.

6.3 Recursive sLORETA-FOCUSS

This is a hybrid method for localization purposes. Developed by Rafik et al. [3], recursive sLORETA-FOCUSS works in an iterative way. Hence, it utilizes the features of sLORETA and FOCUSS in a recursive manner to localize the brain sources. The solution is started from a smooth source distribution, which is further carried on using an iterative algorithm to enhance the strength of prominent elements and consequently reducing the strength for nonprominent sources in the solution. It implies that sLORETA-FOCUSS suppresses the sources that have current density close to zero and recognizes the solution with higher current density only.

Initially, the current density is estimated using the sLORETA method. Hence, numerically, the current density is given by

$$J_{\text{sLORETA}} = \hat{S}_j \times J_{\text{MNE}} \tag{6.12}$$

where $\hat{S}_j$ is the variance of the estimated current density, and J_{MNE} is the estimated current density related to minimum norm estimation. After this step, the initial value for the weight matrix is calculated using the current density obtained through the sLORETA technique. Thus, the initial value of weight matrix W is given by

$$W_0 = diag[J_{sLORETA}(i)] \tag{6.13}$$

Following this step, the current density distribution is estimated using the following equation:

$$\hat{J}_i = W_i W_i^T L^T (L W_i W_i^T L^T)^+ Y \tag{6.14}$$

For each step, the weight matrix is updated, which makes the algorithm iterative in nature. The matrix is updated using the following equation:

$$W_i = P W_{i-1} \{ diag[\hat{J}_{i-1}(1), \hat{J}_{i-1}(2), \dots, \hat{J}_{i-1}(3M)] \} \tag{6.15}$$

where $\hat{J}_{i-1}(n)$ is the nth element of vector $\hat{J}$ at the $(i-1)$th iteration. P is a diagonal matrix, which is given as follows:

$$P = diag \left[1 / \|L_1\|, 1 / \|L_2\|, \dots, 1 / \|L_{3M}\| \right] \tag{6.16}$$

This procedure is repeated until the solution no longer changes.

It should be noted that this update of the weight matrix is carried out by FOCUSS, which is a recursive technique. Thus, the weight matrix is altered in an iterative manner based on the data provided by the current density estimates of the previous ith iteration. The process is repeated (and so the name *recursive*) to eradicate the nonactive areas of the brain. Thus, after the elimination, new space is defined only for the active area. These steps are repeated until convergence. The convergence here defines the number of nodes in the newly defined solution space as less than the number of sensors used for measurements.

The technique was analyzed by simulating two current dipoles using MATLAB, and a comparison is formed between various localization algorithms such as sLORETA, FOCUSS, sLORETA-FOCUSS, and recursive sLORETA-FOCUSS. According to the simulated images, sLORETA produced smooth and diffused reconstructed images for two dipoles, which show the inability of sLORETA to localize the dipoles correctly. The FOCUSS technique alone provides sparse solution, which does not sufficiently solve the need for a method to provide satisfactory localization results. The hybrid sLORETA-FOCUSS has exact convergence to the dipole with no localization error. However, the problem with this hybrid technique is the generation of small replica sources besides the space solution. The recursive sLORETA-FOCUSS technique provides the same result as the simulated dipole. The computational time taken by the newly designed hybrid technique—that is, recursive sLORETA-FOCUSS—is lesser, as it takes only 323.7031 seconds unlike the sLORETA-FOCUSS (330.4531 seconds) and FOCUSS (494.0313 seconds). The extra time taken by FOCUSS is due to its recursive nature. However, sLORETA alone provides a solution with low resolution.

6.4 Shrinking LORETA-FOCUSS

This is another hybrid technique that takes advantage of the LORETA and FOCUSS techniques in a hybrid way. The technique was developed and discussed by He Sheng [4]. To understand the idea for shrinking LORETA-FOCUSS, the major steps for LORETA-FOCUSS are defined first. Hence, the main idea for LORETA-FOCUSS is to first compute the estimated solution for current density—that is, J'_{LORETA}. After this step, the weight matrix W is constructed using the following equation:

$$W_i = PW_{i-1}(diag(J'_{1(i-1)}, J'_{2(i-1)}, \ldots, J'_{3M(i-1)}))$$ (6.17)

where $J_{n(i-1)}$ is the nth element of vector $\mathbf{J}'$ at $(i-1)$th iteration. However, P is the diagonal matrix for deeper source compensation and is given as:

$$P = diag\left[1/\|L_1\|, 1/\|L_2\|, \ldots, 1/\|L_{3M}\|\right]$$ (6.18)

The current density is computed after this step, which is given by the following equation:

$$J_i = W_i W_i^T L^T (\mathbf{L} W_i W_i^T L^T)^{-1} \mathbf{Y}$$ (6.19)

These steps are continued until the convergence of a solution.

Shrinking LORETA-FOCUSS provides a novel idea of search space reduction for each iteration for the solution, which results in time reduction. The search space is often defined around the nodes with prominent current strength, and they compose the solution space for the next iteration. Hence, to avoid the error accumulation, the solution in each iteration is readjusted before it affects the weighting matrix of the next step. It should be noted that during FOCUSS iteration, some of the meaningful nodes are also eliminated. Because the weighting matrices are normally constructed by the previous estimation, these nodes cannot be taken back in the subsequent steps as they are always zero weighted. An appropriate way to solve such a problem is to smooth the estimated topography after each iteration. Hence, for this algorithm, a smoothing operator is defined as matrix L, such that the smoothed topography is given by

$$L \cdot \hat{\mathbf{J}} = l_1^T, l_2^T, \ldots, l_M^T \tag{6.20}$$

For a regular cubic grid of nodes with a minimum internode distance of d, the smoothed current densities are

$$l_i = \frac{1}{s_i + 1}\left(\hat{J}_i + \sum_u \hat{J}_u\right) \tag{6.21}$$

Under the constraint of

$$\left\|r_i - r_u\right\| \le d \tag{6.22}$$

Here, r_i denotes the position vector of ith node, and s_i denotes the number of neighboring nodes within the region defined by u.

Hence, shrinking LORETA-FOCUSS estimates the current density and constructs the weight matrices as defined earlier. However, with the help of a smoothing operator (defined earlier), the search space is reduced by retaining only the prominent nodes and discarding the weak nodes. This procedure will eventually reduce the size of current density matrix **J** and columns of leadfield matrix **L**. The process is repeated until convergence, and the solution of the last iteration before smoothing is considered as the final solution. The results discussed demonstrate that the technique provides reconstruction of sources with relatively high spatial resolution as compared with the LORETA algorithm. The localization capability is compared with other algorithms in terms of energy error (E_{enrg}), which is computed as follows:

$$E_{enrg} = 1 - \frac{\left\|\hat{J}_{max}\right\|}{\left\|J_{simu}\right\|} \tag{6.23}$$

where $\left\|\hat{J}_{max}\right\|$ is the power of maxima in the estimated current density, and $\left\|J_{simu}\right\|$ is the power of the simulated point source.

The results demonstrate that the mean localization error for this technique is low (0.72) as compared with the LORETA (13.41) and LORETA-FOCUSS (2.33) algorithms. However, the energy error as defined above is also numerically smaller (0.73) when compared with LORETA (96.75) and LORETA-FOCUSS (8.44). This method is evaluated on simulated data only, and the algorithm is not validated using experimental data. Table 6.1 shows

Table 6.1 Comparison between Various Techniques[a]

	LORETA	LORETA-FOCUSS	Shrinking LORETA-FOCUSS
E_{loc} (mm)	13.41	2.33	0.72
E_{max_loc} (mm)	59.81	38.34	35.69
$E_{loc} \leq 7$ (mm)	400 nodes	2136 nodes	2307 nodes
$E_{loc} \leq 14$ (mm)	1591 nodes	2330 nodes	2379 nodes
$\bar{E}_{enrg}$ (%)	96.75	8.44	0.73
E_{max_enrg} (%)	99.76	76.54	79.98
$E_{enrg} \leq 0.01\%$	0 nodes	1292 nodes	2109 nodes
$E_{enrg} \leq 1\%$	0 nodes	1729 nodes	2207 nodes

[a] LORETA, Low-resolution brain electromagnetic tomography; LORETA-FOCUSS, low-resolution brain electromagnetic tomography–focal underdetermined system solution.

the comparison between various techniques in terms of energy error [4]. A comparison of the localization ability for LORETA, LORETA-FOCUSS, and shrinking LORETA-FOCUSS is presented in the following section.

6.5 *Standardized shrinking LORETA-FOCUSS*

This is another hybrid technique that takes into consideration the formulation provided by both LORETA and FOCUSS (standardized versions). This technique was introduced and explained by Hesheng et al. [5]. This technique makes use of the recursive procedure, which is initialized by the smooth solution provided by sLORETA. Hence, the reweighted minimum norm is introduced by FOCUSS. Furthermore, an important feature—that is, standardization—is involved in the recursive process for enhancement of localization capability. The technique is improved further by adjustment of the source space automatically to the estimate of the previous step and by the inclusion of temporal information.

The technique starts by estimating the current density using the sLORETA formulation as described in Pascual-Marqui [6]. Mathematically, it is given by

$$\mathbf{J}' = T\mathbf{L}\mathbf{J} = L^T (LL^T + \alpha \mathbf{H})^+ \mathbf{L}\mathbf{J} = S_{J'}J \qquad (6.24)$$

After this, the weight matrix is initialized as:

$$W_0 = diag[\mathbf{J}'(1), \mathbf{J}'(2), \dots, J'(3M)] \qquad (6.25)$$

The next step is to calculate the source power using the standardized FOCUSS formulation. For this, the following set of equations is used:

$$J_i' = W_i W_i^T L^T (L W_i W_i^T L^T)^+ L J = R_i J \tag{6.26}$$

where $R_i \in \Re^{3 \times 3}$ is the resolution matrix and is defined as:

$$R_i = W_i W_i^T L^T (L W_i W_i^T L^T)^+ L \tag{6.27}$$

Only prominent nodes with maximum strength are retained as was defined in the shrinking technique. After this step, the solution space is redefined with only those nodes having significant strength. The reduction of solution space through selection of prominent nodes is application specific. With the new solution space, new matrices for current density (**J**) and leadfield (**L**) are defined, which have values corresponding to prominent nodes only. The weight matrix is now subject to updates according to the following equation:

$$W_i = P W_{i-1} \{ diag[J_{1(i-1)}', J_{2(i-1)}', \ldots, J_{3M(i-1)}'] \} \tag{6.28}$$

This procedure is iterative in nature, which implies that it is repeated until convergence. Thus, the solution of the last iteration before smoothing is the final solution. If the solution remains the same for two consecutive steps, then the iterations are stopped. In addition, if the solution of any iteration is less sparse than the solution estimated by the previous iteration, then the iterations are stopped.

This technique was validated using forward modeling with spherical and realistic head modeling. The leadfield matrix was calculated using the finite element method. The comparison of four different methods, which are WMN, sLORETA, FOCUSS, and standardized shrinking LORETA-FOCUSS (SSLOFO), was produced in terms of localization error and localization ability with noise-free simulations.

According to the table, it is clear that SSLOFO has the least localization error and is more efficient in source localization as compared with sLORETA, FOCUSS, and WMN. The same results were observed for noisy data, where the correlation coefficient between the simulated wave and reconstructed wave for SSLOFO was significantly high as compared with the mentioned techniques. Hence, SSLOFO may be considered as best among the classical techniques.

Summary

This chapter dealt with the hybrid techniques that were developed by mixing one of the classical techniques with another to maximize the localization capability and reduce the error. Hence, the first hybrid WMN

Table 6.2 Localization Capability Comparison between Various
Inverse Techniques[a]

	WMN	sLORETA	FOCUSS	SSLOFO
E_{loc} (mm)	20.05	0	2.33	0
E_{max_loc} (mm)	81.03	0	38.34	0
STD of localization errors (mm)	12.57	0	4.50	0
$\bar{E}_{enrg}$ (%)	96.16	99.55	8.44	2.99
E_{max_enrg} (%)	99.78	99.85	76.55	40.78
STD of energy errors (%)	3.37	0.21	20.62	5.36

Source: The table is reproduced from L. Hesheng et al., *IEEE Transactions on Biomedical Engineering*, vol. 52, pp. 1681–1691, 2005.

[a]   FOCUSS, Focal underdetermined system solution; sLORETA, standardized low-resolution brain electromagnetic tomography; SSLOFO, standardized shrinking low-resolution brain electromagnetic tomography–focal underdetermined system solution; STD, standard deviation; WMN, weighted minimum norm.

method was discussed with its mathematical derivations and results. Then, WMN–LORETA was discussed with its basic formulations and results were obtained. Table 6.2 shows comparison between various techniques in terms of energy error [5]. The discussion was continued for the iterative method based on hybridization of sLORETA and FOCUSS—that is, recursive sLORETA-FOCUSS. Finally, shrinking LORETA-FOCUSS and its advanced version, SSLOFO, were discussed with their major steps and results were obtained. Hence, it is concluded that by mixing various classical techniques, better estimation may be obtained. However, SSLOFO performed best among these hybrid techniques as it has the least localization error and maximum correlation with simulated data with and without noise condition.

References

1. C. Y. Song, Q. Wu, and T. G. Zhuang, Hybrid Weighted Minimum Norm Method A new method based LORETA to solve EEG inverse problem, in *27th Annual International Conference of the Engineering in Medicine and Biology Society, 2005 (IEEE-EMBS 2005)*, New York: IEEE, pp. 1079–1082, 2005.
2. K. Rafik, W. Zouch, A. Taleb-Ahmed, and A. B. Hamida, A new combining approach to localizing the EEG activity in the brain WMN and LORETA solution. *International Conference on BioMedical Engineering and Informatics, 2008 (BMEI 2008)*, Vol. 1, New York: IEEE, pp. 821–824, 2008.
3. K. Rafik, B. H. Ahmed, F. Imed, and T.-A. Abdelmalik, Recursive sLORETA-FOCUSS algorithm for EEG dipoles localization, in *2008 First International*

Workshops on Image Processing Theory, Tools and Applications, 2008 (IPTA 2008), New York: IEEE, pp. 1–5, 2008.

4. L. He Sheng, Y. Fusheng, G. Xiaorong, and G. Shangkai, Shrinking LORETA-FOCUSS: A recursive approach to estimating high spatial resolution electrical activity in the brain, in *First International IEEE EMBS Conference on Neural Engineering, 2003*, New York: IEEE, pp. 545–548, 2003.

5. L. Hesheng, P. H. Schimpf, D. Guoya, G. Xiaorong, Y. Fusheng, and G. Shangkai, Standardized shrinking LORETA-FOCUSS (SSLOFO): A new algorithm for spatio-temporal EEG source reconstruction, *IEEE Transactions on Biomedical Engineering*, vol. 52, pp. 1681–1691, 2005.

6. R. Pascual-Marqui, Standardized low-resolution brain electromagnetic tomography (sLORETA): Technical details, *Methods and Findings in Experimental and Clinical Pharmacology*, vol. 24, pp. 5–12, 2002.

chapter seven

EEG inverse problem III
Subspace-based techniques

Introduction

Over the past few decades, a variety of techniques have been developed for brain source localization using noninvasive measurements of brain activities, such as EEG and magnetoencephalography (MEG). Brain source localization uses measurements of the voltage potential or magnetic field at various locations on the scalp and then estimates the current sources inside the brain that best fit these data using different estimators.

The earliest efforts to quantify the locations of the active EEG sources in the brain occurred more than 50 years ago when researchers began to relate their electrophysiological knowledge about the brain to the basic principles of volume currents in a conductive medium [1–3]. The basic principle is that an active current source in a finite conductive medium produces volume currents throughout the medium, which lead to potential differences on its surface. Given the special structure of the pyramidal cells in the cortical area, if enough of these cells are in synchrony, volume currents large enough to produce measurable potential differences on the scalp will be generated.

The process of calculating scalp potentials from current sources inside the brain is generally called the forward problem. If the locations of the current sources in the brain are known and the conductive properties of the tissues within the volume of the head are also known, the potentials on the scalp can be calculated from the electromagnetic field principles. Conversely, the process of estimating the locations of the sources of the EEG from measurements of the scalp potentials is called the inverse problem.

Source localization is an inverse problem, where a unique relationship between the scalp-recorded EEG and neural sources may not exist. Therefore, different source models have been investigated. However, it is well established that neural activity can be modeled using equivalent current dipole models to represent well-localized activated neural sources [4,5].

Numerous studies have demonstrated a number of applications of dipole source localization in clinical medicine and neuroscience research, and many algorithms have been developed to estimate dipole locations

[6,7]. Among the dipole source localization algorithms, the subspace-based methods have received considerable attention because of their ability to accurately locate multiple closely spaced dipole sources and/or correlated dipoles. In principle, subspace-based methods find (maximum) peak locations of their cost functions as source locations by employing certain projections onto the estimated signal subspace, or alternatively, onto the estimated noise-only subspace (the orthogonal complement of the estimated signal subspace), which are obtained from the measured EEG data. The subspace methods that have been studied for MEG/EEG include classic multiple signal classification (MUSIC) [8] and recursive types of MUSIC: for example, recursive-MUSIC (R-MUSIC) [6] and recursively applied and projected-MUSIC (RAP-MUSIC) [6]. Mosher et al. [4] pioneered the investigation of MEG source dipole localization by adapting the MUSIC algorithm, which was initially developed for radar and sonar applications [8]. Their work has made an influential impact on the field, and MUSIC has become one of most popular approaches in MEG/EEG source localization. Extensive studies in radar and sonar have shown that MUSIC typically provides biased estimates when sources are weak or highly correlated [9]. Therefore, other subspace algorithms that do not provide large estimation bias may outperform MUSIC in the case of weak and/or correlated dipole sources when applied to dipole source localization. In 1999, Mosher and Leahy [6] introduced RAP-MUSIC. It was demonstrated in one-dimensional (1D) linear array simulations that when sources were highly correlated, RAP-MUSIC had better source resolvability and smaller root mean-squared error of location estimates as compared with classic MUSIC.

In 2003, Xu et al. [10] proposed a new approach to EEG three-dimensional (3D) dipole source localization using a nonrecursive subspace algorithm called first principle vectors (FINES). In estimating source dipole locations, the present approach employs projections onto a subspace spanned by a small set of particular vectors (FINES vector set) in the estimated noise-only subspace instead of the entire estimated noise-only subspace in the case of classic MUSIC. The subspace spanned by this vector set is, in the sense of principal angle, closest to the subspace spanned by the array manifold associated with a particular brain region. By incorporating knowledge of the array manifold in identifying FINES vector sets in the estimated noise-only subspace for different brain regions, the present approach is able to estimate sources with enhanced accuracy and spatial resolution, thus enhancing the capability of resolving closely spaced sources and reducing estimation errors.

In this chapter, we outline the MUSIC and its variant, the RAP-MUSIC algorithm, and the FINES as representatives of the subspace techniques in solving the inverse problem with brain source localization.

Because we are primarily interested in the EEG/MEG source localization problem, we have restricted our attention to methods that do not impose specific constraints on the form of the array manifold. For this reason, we do not consider methods such as estimation of signal parameters via rotational invariance techniques (ESPRIT) [11] or root multiple signal classification-MUSIC (ROOT-MUSIC), which exploits shift invariance or Vandermonde structure in specialized arrays.

Subspace methods have been widely used in applications related to the problem of direction of arrival estimation of far-field narrowband sources using linear arrays. Recently, subspace methods started to play an important role in solving the issue of localization of equivalent current dipoles in the human brain from measurements of scalp potentials or magnetic fields, namely, EEG or MEG signals [6]. These current dipoles represent the foci of neural current sources in the cerebral cortex associated with neural activity in response to sensory, motor, or cognitive stimuli. In this case, the current dipoles have three unknown location parameters and an unknown dipole orientation. A direct search for the location and orientation of multiple sources involves solving a highly nonconvex optimization problem.

One of the various approaches that can be used to solve this problem is the MUSIC [8] algorithm. The main attractions of MUSIC are that it can provide computational advantages over least squares methods in which all sources are located simultaneously. Moreover, they search over the parameter space for each source, avoiding the local minima problem, which can be faced while searching for multiple sources over a nonconvex error surface. However, two problems related to MUSIC implementation often arise in practice. The first one is related to the errors in estimating the signal subspace, which can make it difficult to differentiate "true" from "false" peaks. The second is related to the difficulty in finding several local maxima in the MUSIC algorithm because of the increased dimension of the source space. To overcome these problems, the RAP-MUSIC and FINES algorithms were introduced.

In the remaining part of this chapter, the fundamentals of matrix subspaces and related theorems in linear algebra are first outlined. Next, the EEG forward problem is briefly described, followed with a detailed discussion of the MUSIC, the RAP-MUSIC, and the FINES algorithms.

7.1 Fundamentals of matrix subspaces

7.1.1 Vector subspace

Consider a set of vectors S in the n-dimension real space $\mathbf{R}^n$.

S is a subspace of $\mathbf{R}^n$ if it satisfies the following properties:

- The zero vector ϵ S.
- S is closed under addition. This means that if $\mathbf{u}$ and $\mathbf{v}$ are vectors in S, then their sum $\mathbf{u} + \mathbf{v}$ must be in S.
- S is closed under scalar multiplication. This means that if $\mathbf{u}$ is a vector in $\mathbf{H}$ and c is any scalar, the product $c\mathbf{u}$ must be in S.

7.1.2 Linear independence and span of vectors

Vectors $\mathbf{a}_1, \mathbf{a}_2, \ldots, \mathbf{a}_n \in \mathbf{R}^m$ are linearly independent if none of them can be written as a linear combination of the others:

$$\sum_{j=1}^{n} \alpha_j a_j = 0 \quad \text{implies} \quad \alpha(1{:}n) = 0 \tag{7.1}$$

Given $\mathbf{a}_1, \mathbf{a}_2, \ldots, \mathbf{a}_n \in \mathbf{R}^m$, the set of all linear combinations of these vectors is a subspace $S \in \mathbf{R}^m$:

$$S = span\{\mathbf{a}_1, \mathbf{a}_2, \ldots, \mathbf{a}_n\} \tag{7.2}$$

7.1.3 Maximal set and basis of subspace

If the set $\phi = \{\mathbf{a}_1, \mathbf{a}_2, \ldots, \mathbf{a}_n\}$ represents the maximum number of independent vectors in $\mathbf{R}^m$, then it is called the maximal set.

If the set of vectors $\varphi = \{\mathbf{a}_1, \mathbf{a}_2, \ldots, \mathbf{a}_k\}$ is a maximal set of subspace S, then $S = span\{\mathbf{a}_1, \mathbf{a}_2, \ldots, \mathbf{a}_k\}$ and φ is called the basis of S.

If S is a subspace of $\mathbf{R}^m$, then it is possible to find various bases of S. All bases for S should have the same number of vectors (k).

The number of vectors in the bases (k) is called the **dimension** of the subspace and denoted as $k = \dim(S)$.

7.1.4 The four fundamental subspaces of $\mathbf{A} \in \mathbf{R}^{m \times n}$

Matrix $\mathbf{A} \in \mathbf{R}^{m \times n}$ has four fundamental subspaces defined as follows.

The column space of $\mathbf{A}$ is defined as:

$$C(\mathbf{A}) = span\{\mathbf{c}_1, \cdots, \mathbf{c}_n\}$$
$$C(\mathbf{A}) \in \mathbf{R}^m \tag{7.3}$$

The nullspace of $\mathbf{A}$ is defined as:

$$N(\mathbf{A}) = \{\mathbf{x} \in \mathbf{R}^n : \mathbf{A}\mathbf{x} = 0\}$$
$$N(\mathbf{A}) \in \mathbf{R}^n \tag{7.4}$$

The column space of $\mathbf{A}^\mathsf{T}$ is defined as:

$$C(\mathbf{A}^\mathsf{T}) = span\{\mathbf{r}_1, \ldots, \mathbf{r}_m\}$$
$$C(\mathbf{A}^\mathsf{T}) \in \mathbf{R}^n$$

(7.5)

The nullspace of $\mathbf{A}^\mathsf{T}$ is defined as:

$$N(\mathbf{A}^\mathsf{T}) = \{\mathbf{y} \in \mathbf{R}^m : \mathbf{A}^\mathsf{T}\mathbf{y} = 0\}$$
$$N(\mathbf{A}^\mathsf{T}) \in \mathbf{R}^m$$

(7.6)

The column space and row space have equal dimension $r = rank(\mathbf{A})$. The nullspace $N(\mathbf{A})$ has the dimension $n - r$, $N(\mathbf{A}^\mathsf{T})$ has the dimension $m - r$, and the dimensions of the four fundamental subspaces of matrix $\mathbf{A}$ ϵ $\mathbf{R}^{m \times n}$ are given as follows:

$$\begin{aligned} \dim[C(\mathbf{A})] + \dim[N(\mathbf{A})] = \\ (r) \quad + \quad (n-r) \quad\quad = n \\ \dim[C(\mathbf{A}^\mathsf{T})] + \dim[N(\mathbf{A}^\mathsf{T})] = \\ (r) \quad\quad + \quad (m-r) \quad\quad = m \end{aligned}$$

(7.7)

The row space $C(\mathbf{A}^\mathsf{T})$ and nullspace $N(\mathbf{A})$ are orthogonal complements (Figure 7.1). The orthogonality comes directly from the equation $\mathbf{A}x = 0$.

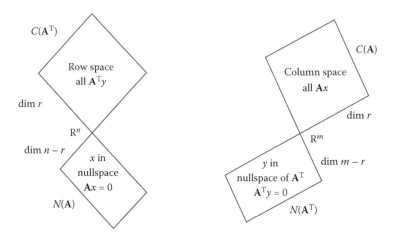

Figure 7.1 Dimensions and orthogonality for any m by n matrix $\mathbf{A}$ of rank r [12].

Each x in the $N(A)$ is orthogonal to all the rows of A as shown in the following equation:

$$Ax = 0$$

$$\begin{bmatrix} (\text{row } 1) \\ \cdots \\ (\text{row } m) \end{bmatrix} \begin{bmatrix} \\ x \\ \end{bmatrix} = \begin{bmatrix} 0 \\ 0 \\ \vdots \\ 0 \end{bmatrix} \begin{matrix} \leftarrow x \text{ is orthogonal to row } 1 \\ \\ \\ \leftarrow x \text{ is orthogonal to row } m \end{matrix} \tag{7.8}$$

The column space $C(A)$ and nullspace of A^T are orthogonal complements.

The orthogonality comes directly from the equation $A^Ty = 0$.

Each y in the nullspace of A^T is orthogonal to all the columns of A as shown in the following equation:

$$A^Ty = 0$$

$$\begin{bmatrix} (\text{column } 1) \\ \cdots \\ (\text{column } n) \end{bmatrix} \begin{bmatrix} \\ y \\ \end{bmatrix} = \begin{bmatrix} 0 \\ 0 \\ \vdots \\ 0 \end{bmatrix} \begin{matrix} \leftarrow y \text{ is orthogonal to column } 1 \\ \\ \\ \leftarrow y \text{ is orthogonal to column } n \end{matrix} \tag{7.9}$$

7.1.5 Orthogonal and orthonormal vectors

Subspaces $S_1, ..., S_p$ in R^m are mutually orthogonal if

$$x^Ty = 0 \quad \text{whenever } x \in S_i \text{ and } y \in S_j \text{ for } i \neq j \tag{7.10}$$

Orthogonal complement subspace in R^m is defined by

$$S^\perp = \{y \in R^m : y^Tx = 0 \quad \text{for all } x \in S\} \tag{7.11}$$

$$\dim(S) + \dim(S^\perp) = m \tag{7.12}$$

Matrix $Q \in R^{m \times m}$ is said to be orthogonal if $Q^T Q = I_{m \times m}$, where I is the identity matrix.

If $Q = \begin{bmatrix} q_1 & q_2 & \cdots & q_m \end{bmatrix}$ is orthogonal, then q_i forms the orthonormal basis for R^m.

Theorem [12]: If $\mathbf{V}_1 \in \mathbf{R}^{m \times r}$ has orthogonal column vectors, then there exists $\mathbf{V}_2 \in \mathbf{R}^{m \times (m-r)}$ such that $C(\mathbf{V}_1)$ and $C(\mathbf{V}_2)$ are orthogonal.

7.1.6 Singular value decomposition

Singular value decomposition (SVD) is a useful tool in handling the problem of orthogonality. SVD deals with orthogonality through its intelligent handling of the matrix rank problem.

Theorem [12]: If $\mathbf{A}$ is a full rank real m-by-n matrix and $m > n$, then there exists orthogonal matrices:

$$\mathbf{U} = [\mathbf{u}_1, \cdots, \mathbf{u}_m] \in \mathbf{R}^{m \times m}$$
$$\mathbf{V} = [\mathbf{v}_1, \cdots, \mathbf{v}_n] \in \mathbf{R}^{n \times n}$$
$$\mathbf{U}^T \mathbf{A} \mathbf{V} = \Sigma = diag(\sigma_1, ..., \sigma_n) \in \mathbf{R}^{m \times n} \text{ where } \sigma_1 \geq \sigma_2 \geq \cdots \geq \sigma_n \geq 0 \quad (7.13)$$

$$\mathbf{A} = \mathbf{U} \Sigma \mathbf{V}^T = \sum_{i=1}^{n} \sigma_i \mathbf{u}_i \mathbf{v}_i^T$$

where σ_i is the singular value of $\mathbf{A}$, and $\mathbf{u}_i$ and $\mathbf{v}_i$ are the left and right singular vectors, respectively.

It is easy to verify that $\mathbf{A}\mathbf{V} = \mathbf{U}\Sigma$ and $\mathbf{A}^T\mathbf{U} = \mathbf{V}\Sigma$, where

$$\left.\begin{array}{r}\mathbf{A}\mathbf{v}_i = \sigma_i \mathbf{u}_i \\ \mathbf{A}^T\mathbf{u}_i = \sigma_i \mathbf{v}_i\end{array}\right\} \quad i = 1 : \min\{m, n\} \quad (7.14)$$

If $\mathbf{A}$ is rank deficient with $\text{rank}(\mathbf{A}) = r$, then

$$\sigma_1 \geq \sigma_2 \geq \cdots \geq \sigma_r > \sigma_{r+1} = \cdots = 0 \quad (7.15)$$

$$\begin{array}{ll} C(\mathbf{A}) = \text{span}\{\mathbf{u}_1, ..., \mathbf{u}_r\} \in \mathbf{R}^m, & N(\mathbf{A}) = \text{span}\{\mathbf{v}_{r+1}, ..., \mathbf{v}_n\} \in \mathbf{R}^n \\ C(\mathbf{A}^T) = \text{span}\{\mathbf{v}_1, ..., \mathbf{v}_r\} \in \mathbf{R}^n, & N(\mathbf{A}^T) = \text{span}\{\mathbf{u}_{r+1}, ..., \mathbf{u}_m\} \in \mathbf{R}^m \end{array} \quad (7.16)$$

7.1.7 Orthogonal projections and SVD

Let S be a subspace of $\mathbf{R}^n$. $\mathbf{P} \in \mathbf{R}^{n \times n}$ is the orthogonal projection onto S if $\mathbf{P}x \in S$ for each vector x in $\mathbf{R}^n$.

The projection matrix $\mathbf{P}$ satisfies two properties:

$$\mathbf{P}^2 = \mathbf{P}$$
$$\mathbf{P}^T = \mathbf{P} \quad (7.17)$$

From this definition, if $x \in R^n$, then $\mathbf{P}x \in S$ and $(\mathbf{I} - \mathbf{P})x \in S^{\perp}$.

Suppose $\mathbf{A} = \mathbf{U\Sigma V}^T$ is the SVD of $\mathbf{A}$ of rank r and

$$\mathbf{U} = \underset{\substack{r \quad m-r}}{[\mathbf{U}_r \quad \tilde{\mathbf{U}}_r]} \quad \mathbf{V} = \underset{\substack{r \quad n-r}}{[\mathbf{V}_r \quad \tilde{\mathbf{V}}_r]} \tag{7.18}$$

There are several important orthogonal projections associated with SVD:

$$\begin{aligned}
\mathbf{V}_r\mathbf{V}_r^T &= \text{projection of } x \in R^n \text{ onto } C(\mathbf{A}^T) \\
\tilde{\mathbf{V}}_r\tilde{\mathbf{V}}_r^T &= \text{projection of } x \in R^n \text{ onto } C(\mathbf{A}^T)^{\perp} = N(\mathbf{A}) \\
\mathbf{U}_r\mathbf{U}_r^T &= \text{projection of } x \in R^m \text{ onto } C(\mathbf{A}) \\
\tilde{\mathbf{U}}_r\tilde{\mathbf{U}}_r^T &= \text{projection of } x \in R^m \text{ onto } C(\mathbf{A})^{\perp} = N(\mathbf{A}^T)
\end{aligned} \tag{7.19}$$

7.1.8 Oriented energy and the fundamental subspaces

Define the unit ball (UB) in R^m as:

$$\text{UB} = \left\{ \mathbf{q} \in R^m \mid \|\mathbf{q}\|_2 = 1 \right\} \tag{7.20}$$

Let $\mathbf{A}$ be an $m \times n$ matrix, then for any unit vector $\mathbf{q} \in R^m$; the energy E_q measured in direction $\mathbf{q}$ is defined as:

$$E_q[\mathbf{A}] = \sum_{k=1}^{n} (\mathbf{q}^T \mathbf{a}_k)^2 \tag{7.21}$$

The energy E_S measured in a subspace $S \subset R^m$. $S \subset R^m$ is defined as follows:

$$E_S[\mathbf{A}] = \sum_{k=1}^{n} \|\mathbf{P}_S(\mathbf{a}_k)\|_2^2 \tag{7.22}$$

where $\mathbf{P}_S(\mathbf{a}_k)$ denotes the orthogonal projection of $\mathbf{a}_k$ onto S.

Theorem: Consider the $m \times n$ matrix $\mathbf{A}$ with its SVD defined as in the SVD theorem, where $m \geq n$, then

$$E_{u_i}[\mathbf{A}] = \sigma_i^2 \tag{7.23}$$

If the matrix **A** is rank deficient with rank $= r$, then there exist directions in $\mathbf{R}^m$ that contain maximum energy and others with minimum and no energy at all.

Proof. For the proof, see [13].

Corollary

$$\max \ E_{q \in UB}[\mathbf{A}] = E_{\mathbf{u}_1}[\mathbf{A}] = \sigma_1^2$$

$$\min \ E_{q \in UB}[\mathbf{A}] = E_{\mathbf{u}_r}[\mathbf{A}] = \sigma_r^2$$

$$E_{q \in UB}[\mathbf{A}] = E_{\mathbf{u}_{r+i}}[\mathbf{A}] = 0 \qquad \text{for} \quad i = 1,2,...,n$$

$$\max \ E_{S \subset \mathbf{R}^m}[\mathbf{A}] = E_{S_{\tilde{u}}^r}[\mathbf{A}] = \sum_{i=1}^{r} \sigma_i^2 \qquad (7.24)$$

$$\min \ E_{S \subset \mathbf{R}^m}[\mathbf{A}] = E_{\left(S_{\tilde{u}}^r\right)^{\perp}}[\mathbf{A}] = \sum_{i=r+1}^{n} \sigma_i^2$$

$$= 0$$

7.1.9 The symmetric eigenvalue problem

Theorem (Symmetric Schur decomposition): If $\mathbf{R} \in \mathbf{R}^{n \times n}$ is symmetric ($\mathbf{A}^T\mathbf{A}$), then there exists an orthogonal $\mathbf{V} \in \mathbf{R}^{n \times n}$ such that

$$\mathbf{V}^T \mathbf{R} \mathbf{V} = \boldsymbol{\Lambda} = \text{diag}(\lambda_1,...,\lambda_n) \qquad (7.25)$$

Moreover, for $k = 1{:}n$, $\mathbf{R}(:,k) = \lambda_k \mathbf{V}(:,k)$.

Proof. For the proof, see Golub and Van Loan [12].

Theorem: If $\mathbf{A} \in \mathbf{R}^{m \times n}$ is symmetric-rank deficient with rank $= r$, then

$$\mathbf{R} = \mathbf{V} \boldsymbol{\Lambda} \mathbf{V}^T = (\mathbf{V}_1 \ \ \mathbf{V}_2) \begin{pmatrix} \boldsymbol{\Lambda}_1 & 0 \\ 0 & 0 \end{pmatrix} \begin{pmatrix} \mathbf{V}_1^T \\ \mathbf{V}_1^T \end{pmatrix}$$

$$\mathbf{V}_1 = (\mathbf{v}_1, \mathbf{v}_2, ..., \mathbf{v}_r) \in \mathbf{R}^{n \times r}$$

$$\mathbf{V}_2 = (\mathbf{v}_{r+1}, \mathbf{v}_{r+2}, ..., \mathbf{v}_n) \qquad (7.26)$$

$$\boldsymbol{\Lambda}_1 = \text{diag}(\lambda_1, \lambda_2, ..., \lambda_r) \in \mathbf{R}^{r \times r}$$

$$\text{span}(\mathbf{v}_1, \mathbf{v}_2, ..., \mathbf{v}_r) = C(\mathbf{R}) = C(\mathbf{R}^T)$$

$$\text{span}(\mathbf{v}_{r+1}, \mathbf{v}_{r+2}, ..., \mathbf{v}_m) = C(\mathbf{R})^{\perp} = N(\mathbf{R}^T) = N(\mathbf{R})$$

Proof. For the proof, see Golub and Van Loan [12].

There are important relationships between SVD of $\mathbf{A} \in \mathbf{R}^{m \times n}$ ($m \geq n$) and Schur decomposition of symmetric matrices $(\mathbf{A}^\mathsf{T}\mathbf{A}) \in \mathbf{R}^{n \times n}$ and $(\mathbf{A}\mathbf{A}^\mathsf{T}) \in \mathbf{R}^{m \times m}$.

If $\mathbf{U}^\mathsf{T}\mathbf{A}\mathbf{V} = \mathrm{diag}(\sigma_1, \ldots, \sigma_n)$ is the SVD of $\mathbf{A}$, then the eigendecomposition of $\mathbf{A}^\mathsf{T}\mathbf{A}$ is

$$\mathbf{V}^\mathsf{T}(\mathbf{A}^\mathsf{T}\mathbf{A})\mathbf{V} = \mathrm{diag}(\sigma_1^2, \ldots, \sigma_n^2) \in \mathbf{R}^{n \times n} \qquad (7.27)$$

and the eigendecomposition of $\mathbf{A}\mathbf{A}^\mathsf{T}$ is

$$\mathbf{U}^\mathsf{T}(\mathbf{A}\mathbf{A}^\mathsf{T})\mathbf{U} = \mathrm{diag}(\sigma_1^2, \ldots, \sigma_n^2, 0, \ldots, 0_m) \in \mathbf{R}^{m \times m} \qquad (7.28)$$

Let the eigendecomposition of **rank r** symmetric correlation matrix $\mathbf{R}_s \in \mathbf{R}^{m \times m}$ be given by

$$\mathbf{R}_s = \mathbf{V}_s\mathbf{\Lambda}_s\mathbf{V}_s^\mathsf{T} = (\mathbf{V}_{s1} \ \ \mathbf{V}_{s2})\begin{pmatrix} \mathbf{\Lambda}_{s1} & 0 \\ 0 & 0 \end{pmatrix}\begin{pmatrix} \mathbf{V}_{s1}^\mathsf{T} \\ \mathbf{V}_{s2}^\mathsf{T} \end{pmatrix}$$

$$\begin{aligned} \mathbf{V}_{s1} &= (\mathbf{v}_1, \mathbf{v}_2, \ldots, \mathbf{v}_r) \in \mathbf{R}^{m \times r} \\ \mathbf{V}_{s2} &= (\mathbf{v}_{r+1}, \mathbf{v}_{r+2}, \ldots, \mathbf{v}_m) \\ \mathbf{\Lambda}_{s1} &= \mathrm{diag}(\lambda_1, \lambda_2, \ldots, \lambda_r) \in \mathbf{R}^{r \times r} \\ \mathrm{span}&(\mathbf{v}_1, \mathbf{v}_2, \ldots, \mathbf{v}_r) = C(\mathbf{R}_s) = C\left(\mathbf{R}_s^\mathsf{T}\right) \\ \mathrm{span}&(\mathbf{v}_{r+1}, \mathbf{v}_{r+2}, \ldots, \mathbf{v}_m) = C(\mathbf{R}_s)^\perp = N\left(\mathbf{R}_s^\mathsf{T}\right) = N\left(\mathbf{R}_s\right) \end{aligned} \qquad (7.29)$$

If the data matrix is corrupted with additive white Gaussian noise of variance σ_{noise}^2, then the eigendecomposition of the full-rank noisy correlation matrix $\mathbf{R}_x = \mathbf{R}_s + \sigma_{noise}^2\mathbf{I}_m$ is given as:

$$\mathbf{R}_x = \mathbf{V}_s\mathbf{\Lambda}_x\mathbf{V}_s^\mathsf{T} = (\mathbf{V}_{s1} \ \ \mathbf{V}_{s2})\begin{pmatrix} \mathbf{\Lambda}_s + \sigma_{noise}^2 & 0 \\ 0 & \sigma_{noise}^2\,\mathbf{I}_{m-r} \end{pmatrix}\begin{pmatrix} \mathbf{V}_{s1}^\mathsf{T} \\ \mathbf{V}_{s2}^\mathsf{T} \end{pmatrix} \qquad (7.30)$$

The eigenvectors $\mathbf{V}_{s1}$ associated with the r largest eigenvalues span the signal subspace or principal subspace. The eigenvectors $\mathbf{V}_{s2}$ associated with the smallest $(m - r)$ eigenvalues, $\mathbf{V}_{s2}$, span the noise subspace.

7.2 The EEG forward problem

The EEG forward problem is simply to find the potential $g(\mathbf{r}, \mathbf{r}_{dip}, \mathbf{d})$ at an electrode positioned on the scalp at a point having position vector $\mathbf{r}$, due to a single dipole with dipole moment $\mathbf{d} = d\mathbf{e}_{dip}$ (with magnitude d and orientation e_{dip}), positioned at r_{dip}. These amounts of scalp potentials can be obtained through the solution of Poisson's equation for different configurations of r_{dip} and d.

For p dipole sources, the electrode potential would be the superposition of their individual potentials:

$$m(\mathbf{r}) = \sum_{i=1}^{p} g(\mathbf{r}, \mathbf{r}_{dip_i}, \mathbf{d}_i) \tag{7.31}$$

This can be rewritten as follows:

$$m(\mathbf{r}) = \sum_{i=1}^{p} g(\mathbf{r}, \mathbf{r}_{dip}) d_i e_i \tag{7.32}$$

where $g(\mathbf{r}, \mathbf{r}_{dip_i})$ has three components in the Cartesian x, y, z directions, and $d_i = (d_{i_x}, d_{i_y}, d_{i_z})$ is a vector consisting of the three dipole magnitude components. As indicated in Equation 7.32, the vector d_i can be written as $d_i e_i$, where d_i is a scalar that represents the dipole magnitude, and e_i is a vector that represents the dipole orientation. In practice, one calculates a potential between an electrode and a reference (which can be another electrode or an average reference).

For p dipoles and L electrodes, Equation 7.32 can be written as:

$$\mathbf{m} = \begin{bmatrix} m_{r_1} \\ \vdots \\ m_{r_L} \end{bmatrix} \begin{bmatrix} g(r_1, r_{dip_1}) & \cdots & g(r_1, r_{dip_p}) \\ \vdots & \ddots & \vdots \\ g(r_L, r_{dip_1}) & \cdots & g(r_L, r_{dip_p}) \end{bmatrix} \begin{bmatrix} d_1 e_1 \\ \vdots \\ d_p e_p \end{bmatrix} \tag{7.33}$$

For L electrodes, p dipoles, and K discrete time samples, the EEG data matrix can be expressed as follows:

$$\mathbf{M} = \begin{bmatrix} m(r_1, 1) & \cdots & m(r_1, K) \\ \vdots & \ddots & \vdots \\ m(r_L, 1) & \cdots & m(r_L, K) \end{bmatrix} = \begin{bmatrix} \mathbf{m}(1) & \cdots & \mathbf{m}(K) \end{bmatrix}$$

$$= \mathbf{G}(r_j, r_{dip_i}) \begin{bmatrix} d(r_1, 1)e_1 & \cdots & d(r_1, K)e_1 \\ \vdots & \ddots & \vdots \\ d(r_L, 1)e_p & \cdots & d(r_L, K)e_p \end{bmatrix} \tag{7.34}$$

$$= \mathbf{G}\big[\mathbf{g}(r_j, r_{dip_i})\big]\mathbf{D}$$

where $\mathbf{m}(k)$ represents the output of array of L electrodes at time k due to p sources (dipoles) distributed over the cerebral cortex, and $\mathbf{D}$ is dipole moments at different time instants.

Each row of the gain matrix $\mathbf{G}\left[\mathbf{g}(r_j, r_{dip_i})\right]$ is often referred to as the leadfield, and it describes the current flow for a given electrode through each dipole position [14].

In the aforementioned formulation, it was assumed that both the magnitude and orientation of the dipoles are unknown. However, based on the fact that apical dendrites producing the measured field are oriented normal to the surface [15], dipoles are often constrained to have such an orientation. In this case, only the magnitude of the dipoles will vary and Equation 7.34 can therefore be rewritten as:

$$
\begin{aligned}
\mathbf{M} &= \begin{bmatrix} g\left(r_1, r_{dip_1}\right)e_1 & \cdots & g\left(r_1, r_{dip_p}\right)e_p \\ \vdots & \ddots & \vdots \\ g\left(r_L, r_{dip_1}\right)e_1 & \cdots & g\left(r_L, r_{dip_p}\right)e_p \end{bmatrix} \begin{bmatrix} d(1,1) & \cdots & d(1,K) \\ \vdots & \ddots & \vdots \\ d(p,1) & \cdots & d(p,K) \end{bmatrix} \\
&= \begin{bmatrix} \mathbf{g}\left(r_j, r_{dip_1}\right)e_1 & \cdots & \mathbf{g}\left(r_j, r_{dip_p}\right)e_p \end{bmatrix}\begin{bmatrix} \mathbf{d}_1 & \cdots & \mathbf{d}_K \end{bmatrix} \\
&= \mathbf{G}\left[\mathbf{g}\left(r_j, r_{dip_i}\right)e_i\right]\mathbf{D}
\end{aligned}
\tag{7.35}
$$

Generally, a noise or perturbation matrix $\mathbf{N}$ is added to the system such that the recorded data matrix $\mathbf{M}$ is given as:

$$
\begin{aligned}
\mathbf{M} &= \begin{bmatrix} \mathbf{m}(1) & \cdots & \mathbf{m}(K) \end{bmatrix} + \begin{bmatrix} \mathbf{n}(1) & \cdots & \mathbf{n}(K) \end{bmatrix} \\
&= \mathbf{G}\left[\mathbf{g}\left(r_j, r_{dip_i}\right)e_i\right]\mathbf{D} + \mathbf{N}
\end{aligned}
\tag{7.36}
$$

where the $L \times K$ noise matrix $\mathbf{N} = [\mathbf{n}(1), ..., \mathbf{n}(K)]$. Under this notation, the inverse problem then consists of finding an estimate $\hat{\mathbf{D}}$ of the dipole magnitude matrix, given the electrode positions and scalp readings $\mathbf{M}$ and using the gain matrix $\mathbf{G}\left[\mathbf{g}(r_j, r_{dip_i})e_i\right]$ calculated in the forward problem.

7.3 The inverse problem

The brain source localization problem based on EEG is termed as EEG source localization or the EEG inverse problem. This problem is ill-posed, because an infinite number of source configurations can produce the same potential at the head surface, and it is underdetermined as the unknown (sources) outnumbers the known (sensors) [11]. In general, the EEG inverse problem estimates the locations, magnitude, and time courses of the neuronal sources that are responsible for the production of potential measured by EEG electrodes.

Various methods were developed to solve the inverse problem of EEG source localization [16]. Among these methods is MUSIC and its variants, RAP-MUSIC and FINES. In the following section, the subspace techniques of MUSIC, RAP-MUSIC, and FINES are outlined and discussed in the context of EEG brain source localization.

7.3.1 The MUSIC algorithm

Consider the leadfield matrix, $G[g(r_j, r_{dip_i})]$, of the p sources and L electrodes as given in Equation 7.4. Assume $G[g(r_j, r_{dip_i})]$ to be of full column rank for any set of distinct source parameters—that is, no array ambiguities exist. The additive noise vector, $n(k)$, is assumed to be zero mean with covariance $NN^T = \delta_n^2 I_L$, where superscript "T" denotes the transpose, I_L is the $L \times L$ identity matrix, and δ_n^2 is the noise variance.

In geometrical language, the measured $m(k)$ vector can be visualized as a vector in L dimensional space. The directional mode vectors $g(r_j, r_{dip_i})e_i$ for $i = 1, 2, ..., p$—that is, the columns of $G[g(r_j, r_{dip_i})e_i]$ state that $m(k)$ is a particular linear combination of the mode vectors; the elements of $d(t)$ are the coefficients of the combination. Note that the $m(k)$ vector is confined to the range space of $G[g(r_j, r_{dip_i})e_i]$. That is, if $G[g(r_j, r_{dip_i})e_i]$ has two columns, the range space is no more than a two-dimensional subspace within the L space, and $m(k)$ necessarily lies in the subspace.

If the data are collected over K samples, then the $L \times L$ covariance matrix of the vector $m(k)$ is given as:

$$R \overset{\Delta}{=} MM^T = G\left[g\left(r_j, r_{dip_i}\right)e_i\right]DD^TG^T\left[g\left(r_j, r_{dip_i}\right)e_i\right] + NN^T \qquad (7.37)$$

or

$$R \overset{\Delta}{=} G\left[g\left(r_j, r_{dip_i}\right)e_i\right]R_sG^T\left[g\left(r_j, r_{dip_i}\right)e_i\right] + \delta_{noise}^2 I_N \qquad (7.38)$$

under the basic assumption that the incident signals and the noise are uncorrelated, and where $M = [m(1)\ m(2), ..., m(K)]$, $D = [d(1)\ d(1), ..., d(K)]$, $N = [n(1)\ n(1), ..., n(K)]$, and $R_s = DD^T$ is the source correlation matrix.

For simplicity, the correlation matrix in Equation 7.38 can be rewritten as:

$$R \overset{\Delta}{=} GR_sG^T + \delta_{noise}^2 I_N \qquad (7.39)$$

where $\mathbf{G} = \mathbf{G}\left[\mathbf{g}(r_j, r_{dip_i})e_i\right]$. Because $\mathbf{G}$ is composed of leadfield vectors, which are linearly independent, the matrix has full rank, and the dipoles correlation matrix $\mathbf{R}_s$ is nonsingular as long as dipole signals are incoherent (not fully correlated). A full rank matrix $\mathbf{G}$ and nonsingular matrix $\mathbf{R}_s$ mean that when the number of dipoles p is less than the number of electrodes L, the $L \times L$ matrix $\mathbf{GR}_s\mathbf{G}^T$ is positive semidefinite with rank p.

Decomposition of the noisy Euclidian space into signal subspace and noise subspace can be performed by applying the eigendecomposition of the correlation matrix of the noisy signal, $\mathbf{R}$. Symmetry simplifies the real eigenvalue problem $\mathbf{Rv} = \lambda\mathbf{v}$ in two ways. It implies that all of $\mathbf{R}$'s eigenvalues λ_i are real and that there is an orthonormal basis of eigenvectors $\mathbf{v}_i$. These properties are the consequence of symmetric real Schur decomposition given in Equation 7.25.

Now, if the covariance matrix $\mathbf{R}$ is noiseless, it is given as:

$$\mathbf{R} \overset{\triangle}{=} \mathbf{GR}_s\mathbf{G}^T \tag{7.40}$$

then the eigendecomposition of the $\mathbf{R}$ as a rank-deficient matrix with rank value equals the number of dipoles (p), which is given as:

$$\mathbf{R} = \mathbf{V}\mathbf{\Lambda}\mathbf{V}^T = (\mathbf{V}_1 \quad \mathbf{V}_2)\begin{pmatrix} \mathbf{\Lambda}_1 & 0 \\ 0 & 0 \end{pmatrix}\begin{pmatrix} \mathbf{V}_1^T \\ \mathbf{V}_2^T \end{pmatrix} \tag{7.41}$$

where $\mathbf{V}_1 = (\mathbf{v}_1, \mathbf{v}_2, ..., \mathbf{v}_p) \in \mathbf{R}^{L\times p}, \mathbf{V}_2 = (\mathbf{v}_{p+1}, \mathbf{v}_{p+2}, ..., \mathbf{v}_n) \in \mathbf{R}^{L\times(L-p)}$, and $\mathbf{\Lambda}_1 = \mathrm{diag}(\lambda_1, \lambda_2, ..., \lambda_p) \in \mathbf{R}^{p\times p}$, where R^p represents the p-dimensional real vector space. The span of the set of vectors in $\mathbf{V}_1$ is the range of matrix $\mathbf{R}$ or $\mathbf{R}^T$, whereas the span of the set of eigenvectors in $\mathbf{V}_2$ is the orthogonal complement of range of $\mathbf{R}$ or its null space. Mathematically, this can be indicated as:

$$\begin{aligned} \mathrm{span}(\mathbf{v}_1, \mathbf{v}_2, ..., \mathbf{v}_r) &= \mathrm{ran}\ (\mathbf{R}) = \mathrm{ran}\ (\mathbf{R}^T) \\ \mathrm{span}(\mathbf{v}_{r+1}, \mathbf{v}_{r+2}, ..., \mathbf{v}_m) &= \mathrm{ran}\ (\mathbf{R})^{\perp} = \mathrm{null}\ (\mathbf{R}^T) = \mathrm{null}\ (\mathbf{R}) \end{aligned} \tag{7.42}$$

If the data matrix is noisy, then its covariance matrix, $\mathbf{R}$, is given as:

$$\mathbf{R} \overset{\triangle}{=} \mathbf{GR}_s\mathbf{G}^T + \mathbf{R}_n \tag{7.43}$$

where $\mathbf{R}_n$ is the noise covariance matrix. If the noise is considered as additive white Gaussian noise, the noise correlation matrix is given as:

$$\mathbf{R}_n = \delta_{noise}^2 \mathbf{I}_N \tag{7.44}$$

Accordingly, $\mathbf{R}_n$ has a single repeated eigenvalue equal to the variance δ_{noise}^2 with multiplicity L, so any vector qualifies as the associated eigenvector, and the eigendecomposition of the noisy covariance matrix in Equation 7.43 is given as:

$$\mathbf{R} = \mathbf{V}\mathbf{\Lambda}\mathbf{V}^T = (\mathbf{V}_1 \quad \mathbf{V}_2) \begin{pmatrix} \mathbf{\Lambda}_1 + \sigma_{noise}^2 & 0 \\ 0 & \sigma_{noise}^2 \mathbf{I}_{L-p} \end{pmatrix} \begin{pmatrix} \mathbf{V}_1^T \\ \mathbf{V}_2^T \end{pmatrix} \tag{7.45}$$

Here, the eigenvectors $\mathbf{V}_1$ associated with the p largest eigenvalues span the *signal subspace* or *principal subspace*.

The eigenvectors $\mathbf{V}_2$ associated with the smallest $(L - p)$ eigenvalues, $\mathbf{V}_2$, span the *noise subspace* or the *null subspace of the matrix* $\mathbf{R}$.

A full rank $\mathbf{G}$ and nonsingular $\mathbf{R}_s$ guarantee that when the number of incident signals p is less than the number of electrodes L, the $L \times L$ matrix $\mathbf{G}\mathbf{R}_s\mathbf{G}^T$ is positive semidefinite with rank p. This means that $L - p$ of its eigenvalues is zero. In this case and as Equation 7.16 indicates, the $N - p$ smallest eigenvalues of $\mathbf{R}$ are equal to δ_{noise}^2 and defining the rank of the matrix becomes a straightforward issue. However, in practice, when the correlation matrix $\mathbf{R}$ is estimated from a finite data sample, there will be no identical values among the smallest eigenvalues. In this case finding the rank of matrix $\mathbf{R}$ becomes a nontrivial problem and can be solved if there is an energy gap between the eigenvalues λ_p and λ_{p+1}—that is, if the ratio $\lambda_{p+1}/\lambda_p < 1$. A gap at p may reflect an underlying rank degeneracy in a matrix $\mathbf{R}$, or simply be a convenient point from which to reduce the dimensionality of a problem. The numerical rank p is often chosen from the statement $\lambda_{p+1}/\lambda_p < 1$.

Now, because $\mathbf{G}$ is full rank, and $\mathbf{R}_s$ is nonsingular, it follows that

$$\mathbf{G}^T \mathbf{v}_i = 0 \quad \text{for} \quad i = d+1, d+2, \ldots, L \tag{7.46}$$

Equation 7.46 implies that a set of eigenvectors that span the noise subspace is orthogonal to the columns of the leadfield matrix, $\mathbf{G}$:

$$\left\{ \mathbf{g}\left(r_j, r_{dip_1}\right) \mathbf{e}_1, \mathbf{g}\left(r_j, r_{dip_2}\right) \mathbf{e}_2, \ldots, \mathbf{g}\left(r_j, r_{dip_p}\right) \mathbf{e}_p \right\} \perp \left\{ \mathbf{v}_{d+1}, \mathbf{v}_{d+2}, \ldots, \mathbf{v}_N \right\} \tag{7.47}$$

Equation 7.47 means that the leadfield vectors corresponding to the locations and orientations of the p dipoles lie in the signal subspace and hence orthogonal to the noise subspace. By searching through all possible leadfield vectors to find those that are perpendicular to the space spanned by the noise subspace eigenvectors of matrix $\mathbf{R}$, the location of the p dipoles can be estimated. This can be accomplished through the principal angles [13] or canonical correlations (cosines of the principal angles).

Let q denote the minimum of the ranks of two matrices, and the canonical or subspace correlation is a vector containing the cosines of the principal angles that reflect the similarity between the subspaces spanned by the columns of the two matrices. The elements of the subspace correlation vector are ranked in decreasing order, and we denote the largest subspace correlation (i.e., the cosine of the smallest principal angle) as:

$$subcorr(\mathbf{A}, \mathbf{B})_1 \tag{7.48}$$

If $subcorr(\mathbf{A}, \mathbf{B})_1 = 1$, then the two subspaces have at least a one-dimensional (1D) subspace in common. Conversely, if $subcorr(\mathbf{A}, \mathbf{B})_1 = 0$, then the two subspaces are orthogonal.

The MUSIC algorithm finds the source locations as those for which the principal angle between the array manifold vector and the noise-only subspace is maximum. Equivalently, the sources are chosen as those that minimize the noise-only subspace correlation $subcorr\left[\mathbf{g}(r_j, r_{dip_i})e_i, \mathbf{V}_2\right]_1$ or maximize the signal subspace correlation $subcorr\left[\mathbf{g}(r_j, r_{dip_i})e_i, \mathbf{V}_1\right]_1$. The square of this signal subspace correlation is given as [17,18]:

$$
\begin{aligned}
subcorr\left[\mathbf{g}(r_j, r_{dip_i})e_i, \mathbf{V}_1\right]_1^2 &= \frac{\left\|\mathbf{P}_S\{\mathbf{g}(r_j, r_{dip_i})e_i\}\right\|^2}{\left\|\{\mathbf{g}(r_j, r_{dip_i})e_i\}\right\|^2} \\
&= \frac{\{\mathbf{g}(r_j, r_{dip_i})e_i\}^H \mathbf{V}_1 \mathbf{V}_1^H \{\mathbf{g}(r_j, r_{dip_i})e_i\}}{\{\mathbf{g}(r_j, r_{dip_i})e_i\}^H \{\mathbf{g}(r_j, r_{dip_i})e_i\}}
\end{aligned}
\tag{7.49}
$$

where $\mathbf{P}_s = \mathbf{V}_1 \mathbf{V}_1^T$ is the projection of the leadfield vectors onto the signal subspace. Theoretically, this function is maximum (one) when $\mathbf{g}(r_j, r_{dip_i})e_i$ corresponds to one of the true locations and orientations of the p dipoles.

Taking into consideration that the estimated leadfield vectors in Equation 7.49 are the product of gain matrix and a polarization or orientation vectors, we can obtain the following:

$$a(\rho, \phi) = g(r_j, r_{dip_i})e_i \tag{7.50}$$

where ρ represents dipole location and ϕ is the dipole orientation; principal angles can be used to represent MUSIC metric for multidimensional leadfield represented by $\mathbf{G}(r_{dip_i}) = \mathbf{g}(r_j, r_{dip_i})$. In this case, MUSIC has to compare space spanned by $\mathbf{G}(r_{dip_i})$ where, $i = 1, 2, \ldots, p$ with the signal subspace spanned by the set of vectors $\mathbf{V}_1$. A similar subspace correlation function to Equation 7.50 can be used to find the *locations of the p dipoles.*

This formula is based on Schmidt's metric for diversely polarized MUSIC, which is given as:

$$subcorr(\mathbf{G}(r_{dip_i}), \mathbf{V}_1)_1^2 = \lambda_{max}(\mathbf{U}_G^H \mathbf{V}_1 \mathbf{V}_1^H \mathbf{U}_G) \tag{7.51}$$

where $\mathbf{U}_G$ contains the left singular vectors of $\mathbf{G}(r_{dip_i})$ and λ_{max} is the maximum eigenvalue of the enclosed expression. The source locations r_{dip_i} can be found as those for which Equation 7.51 is approximately one. The dipoles' orientation is then found from the formula $\mathbf{a}(\rho, \phi) = \mathbf{g}(r_j, r_{dip_i})e_i$.

7.3.2 Recursively applied and projected-multiple signal classification

In MUSIC, errors in the estimate of the signal subspace can make localization of multiple sources difficult (subjective) with regard to distinguishing between "true" and "false" peaks. Moreover, finding several local maxima in the MUSIC metric becomes difficult as the dimension of the source space increases. Problems also arise when the subspace correlation is computed at only a finite set of grid points.

R-MUSIC [19] automates the MUSIC search, extracting the location of the sources through a recursive use of subspace projection. It uses a modified source representation, referred to as the spatiotemporal independent topographies (ITs) model, where a source is defined as one or more nonrotating dipoles with a single time course rather than an individual current dipole. It recursively builds up the IT model and compares this full model to the signal subspace.

In the RAP-MUSIC extension [20,21], each source is found as a global maximizer of a different cost function.

Assuming $\mathbf{g}(r_j, r_{dip_i}, e_i) = \mathbf{g}(r_j, r_{dip_i})e_i$, the first source is found as the source location that maximizes the metric

$$\hat{r}_{dip_1} = arg \ max \left\{ subcorr \left[\mathbf{g}(r_j, r_{dip_i}), \mathbf{V}_1 \right]_1 \right\} \tag{7.52}$$

over the allowed source space, where r is the nonlinear location parameter. The function $subcorr \left[\mathbf{g}(r_j, r_{dip_i}), \mathbf{V}_1 \right]$ is the cosine of the first principal angle between the subspaces spanned by the columns of $\mathbf{g}(r_j, r_{dip_i})$ and $\mathbf{V}_1$ given by

$$subcorr \left[\mathbf{g}(r_j, r_{dip_i}), \mathbf{V}_1 \right]_{dip_1}^2 = \frac{\left[\mathbf{g}(r_j, r_{dip_i})^T \mathbf{V}_1 \mathbf{V}_1^T \mathbf{g}(r_j, r_{dip_i}) \right]}{\mathbf{g}(r_j, r_{dip_i})^T \mathbf{g}(r_j, r_{dip_i})} \tag{7.53}$$

The w-recursion of RAP-MUSIC is given as follows:

$$\hat{r}_{dip_w} = arg \max_{r_{dip}} \left\{ subcorr \left[\Pi^{\perp}_{\hat{G}_{w-1}} g\left(r_j, r_{dip}\right), \Pi^{\perp}_{\hat{G}_{w-1}} \mathbf{V}_1 \right]_1 \right\} \qquad (7.54)$$

where we define

$$\hat{G}_{w-1} = [g(r_j, \hat{r}_{dip_1})\hat{e}_1, \cdots, g(r_j, \hat{r}_{dip_{w-1}})\hat{e}_{w-1}] \qquad (7.55)$$

and

$$\Pi^{\perp}_{\hat{G}_{w-1}} = \left[I - \hat{G}_{w-1}\left(\hat{G}^{\mathrm{T}}_{w-1}\hat{G}_{w-1}\right)^{-1}\hat{G}^{\mathrm{T}}_{w-1} \right] \qquad (7.56)$$

is the projector onto the left-null space of $\hat{G}_{w-1}$. The recursions are stopped once the maximum of the subspace correlation in Equation 7.54 drops below a minimum threshold.

Practical considerations in low-rank E/MEG source localization lead us to prefer the use of the signal rather than the noise-only subspace [22,23]. The development above in terms of the signal subspace is readily modified to computations in terms of the noise-only subspace. Our experience in low-rank forms of MUSIC processing is that the determination of the signal subspace rank need not be precise, as long as the user conservatively overestimates the rank. The additional basis vectors erroneously ascribed to the signal subspace can be considered to be randomly drawn from the noise-only subspace [13]. As described earlier, RAP-MUSIC removes from the signal subspace the subspace associated with each source once it is found. Thus, once the true rank has been exceeded, the subspace correlation between the array manifold and the remaining signal subspace should drop markedly, and thus, additional fictitious sources will not be found.

A key feature of the RAP-MUSIC algorithm is the orthogonal projection operator, which removes the subspace associated with previously located source activity. It uses each successively located source to form an intermediate array gain matrix and projects both the array manifold and the estimated signal subspace into its orthogonal complement, away from the subspace spanned by the sources that have already been found. The MUSIC projection to find the next source is then performed in this reduced subspace.

7.3.3 FINES subspace algorithm

In a recent study by Xu et al. [24], another approach to EEG three-dimensional (3D) dipole source localization using a nonrecursive subspace algorithm, called FINES, has been proposed. The approach employs projections onto a subspace spanned by a small set of particular vectors in the estimated noise-only subspace, instead of the entire estimated noise-only subspace in the case of classic MUSIC. The subspace spanned by this vector set is, in the sense of the principal angle, closest to the subspace spanned by the array manifold associated with a particular brain region. By incorporating knowledge of the array manifold in identifying the FINES vector sets in the estimated noise-only subspace for different brain regions, this approach is claimed to be able to estimate sources with enhanced accuracy and spatial resolution, thus enhancing the capability of resolving closely spaced sources and reducing estimation errors. The simulation results show that, compared with classic MUSIC, FINES has a better resolvability of two closely spaced dipolar sources and also a better estimation accuracy of source locations. In comparison with RAP-MUSIC, the performance of FINES is also better for the cases studied when the noise level is high and/or correlations among dipole sources exist [24].

For FINES, the closeness criterion is the principal angle between two subspaces [12]. FINES identifies a low-dimensional subspace in the noise-only subspace that has the minimum principal angle to the subspace spanned by the section of the leadfield corresponding to a selected location region. In the following section, we describe the FINES algorithms adapted for 3D dipole source localization in EEG.

1. Divide the brain volume into a number of regions of similar volume. For example, a reasonable number of brain regions is 16.
2. For a given region Θ, determine a subspace that well represents the subspace spanned by the leadfield corresponding to the region, that is, $\mathbf{G}(r_{dip_i})$: $r_{dip_i} \in \Theta$. Choose the dimension of this representation subspace as 10 to avoid ambiguity in peak searching and to keep high source resolvability.
3. For a given number of time samples of EEG measurement, form sample correlation matrix $\mathbf{R}$, and then generate the estimated noise-only subspace, that is, the eigenvector matrix $\mathbf{V}_2$.
4. For the given region, identify a set of **10** FINES vectors from the given $\mathbf{V}_2$. The FINES vectors are assumed to be orthonormal.
5. Assume that matrix $\mathbf{V}_{FINES}$ contains the 10 FINES vectors, and search peaks of the following function:

$$J(\rho,\phi) = \frac{1 - a^T(\rho,\phi)\mathbf{V}_{FINES}\mathbf{V}_{FINES}^T a(\rho,\phi)}{\|a(\rho,\phi)\|^2} \tag{7.57}$$

over the selected location region Θ and all possible orientation.

6. Repeat Steps 4 and 5 for other location regions, and p peak locations are the estimates of the p dipoles' location.
7. Similar to MUSIC, instead of maximizing cost function over the six source parameters (three for dipole location and three for dipole orientation), the peak searching can be done over three location parameters only, by minimizing the following:

$$\lambda_{min}\left\{\mathbf{U}_G^T V_{FINES}\mathbf{V}_{FINES}^T\mathbf{U}_G\right\} \tag{7.58}$$

where λ_{min} is the smallest eigenvalue of the bracketed item and the matrix $\mathbf{U}_G$ contains the left singular vectors of $\mathbf{G}(r_{dip_i})$.

Summary

This chapter discussed the subspace concepts in general with the related mathematical derivations. For this, linear independence and orthogonality concepts are discussed with related derivations. For the explanation of the decomposition process for system solution, SVD is explained in detail. Furthermore, the SVD-based algorithms such as MUSIC and RAP-MUSIC are discussed in detail, and then the FINES algorithm is discussed to support the discussion for the subspace-based EEG source localization algorithms.

References

1. R. Plonsey (ed.), *Bioelectric Phenomena*, New York: McGraw-Hill, pp. 304–308, 1969.
2. M. Schneider, A multistage process for computing virtual dipolar sources of EEG discharges from surface information, *IEEE Transactions on Biomedical Engineering*, vol. 19, pp. 1–12, 1972.
3. C. J. Henderson, S. R. Butler, and A. Glass, The localization of the equivalent dipoles of EEG sources by the application of electric field theory, *Electroencephalography and Clinical Neurophysiology*, vol. 39, pp. 117–113, 1975.
4. J. C. Mosher, P. S. Lewis, and R. M. Leahy, Multiple dipole modeling and localization from spatio-temporal MEG data, *IEEE Transactions on Biomedical Engineering*, vol. 39, pp. 541–557, 1992.
5. B. N. Cuffin, EEG Localization accuracy improvements using realistically shaped model, *IEEE Transactions on Biomedical Engineering*, vol. 43(3), pp. 68–71, 1996.
6. J. C. Mosher and R. M. Leahy, Source localization using recursively applied projected (RAP) MUSIC, *IEEE Transactions on Signal Processing*, vol. 74, pp. 332–340, 1999.
7. Y. Kosugi, N. Uemoto, Y. Hayashi, and B. He, Estimation of intra-cranial neural activities by means of regularized neural-network-based inversion techniques, *Neurological Research*, vol. 23(5), pp. 435–446, 2001.

8. R. Schmidt, Multiple emitter location and signal parameter estimation, *IEEE transactions on antennas and propagation*, vol. 34(3), pp. 276–280, 1986.

9. X.-L. Xu and K. M. Buckley, Bias analysis of the MUSIC location estimator, *IEEE Transactions on Signal Processing*, vol. 40(10), pp. 2559–2569, 1992.

10. X.-L. Xu, B. Xu, and B. He, An alternative subspace approach to EEG dipole source localization, *Journal of Physics in Medicine and Biology*, vol. 49(2) pp. 327–343, 2004.

11. R. Roy and T. Kailath, ESPRIT-estimation of signal parameters via rotational invariance techniques, *IEEE Transactions on Acoustics, Speech, and Signal Processing*, vol. 37(7), pp. 984–995, 1989.

12. G. H. Golub and C. F. Van Loan, *Matrix Computations*, 2nd edn, Baltimore, MD: Johns Hopkins University Press, 1984.

13. J. Vandewalle and B. De Moor, A variety of applications of singular value decomposition in identification and signal processing, in *SVD and Signal Processing, Algorithms, Applications, and Architectures*, F. Deprettere (ed.), Amsterdam, The Netherlands: Elsevier, pp. 43–91, 1988.

14. R. D. Pascual-Marqui, Review of methods for solving the EEG inverse problem, *International Journal of Bioelectromagnetism*, vol. 1, pp. 75–86, 1999.

15. A. Dale and M. Sereno, Improved localization of cortical activity by combining EEG and MEG with MRI cortical surface reconstruction: A linear approach, *Journal of Cognitive Neuroscience*, vol. 5(2), pp. 162–176, 1993.

16. R. Grech, T. Cassar, J. Muscat, K. Camilleri, S. Fabri, M. Zervakis, P. Xanthopulos, V. Sakkalis, and B. Vanrumte, Review on solving the inverse problem in EEG source analysis, *Journal of NeuroEngineering and Rehabilitation*, vol. 5(25), pp. 1–13, 2008.

17. R. O. Schmidt, Multiple emitter location and signal parameter estimation, *IEEE Transactions on Antennas and Propagation*, vol. AP-34, pp. 276–280, 1986.

18. R. O. Schmidt, A signal subspace approach to multiple emitter location and spectral estimation, Ph.D. dissertation, Stanford University Stanford, CA, November 1981.

19. J. C. Mosher and R. M. Leahy, Recursive MUSIC: A framework for EEG and MEG source localization, *IEEE Transactions on Biomedical Engineering*, vol. 45(11), pp. 1342–1354, 1998.

20. J. C. Mosher and R. M. Leahy, Source localization using recursively applied and projected (RAP) MUSIC, *IEEE Transactions on Signal Processing*, vol. 47(2), pp. 332–340, 1999.

21. J. J. Ermer, J. C. Mosher, M. Huang, and R. M. Leahy, Paired MEG data set source localization using recursively applied and projected (RAP) MUSIC, *IEEE Transactions on Biomedical Engineering*, vol. 47(9), pp. 1248–1260, 2000.

22. J. C. Mosher and R. M. Leahy, Recursively applied MUSIC: A framework for EEG and MEG source localization, *IEEE Transactions on Biomedical Engineering*, vol. 45, pp. 1342–1354, November 1998.

23. J. C. Mosher and R. M. Leahy, Source localization using recursively applied and projected (RAP) MUSIC, in *Proceedings of the 31st Asilomar Conference on Signals, Systems, and Computers,* New York: IEEE Signal Processing Society, November 2–5, 1997.

24. X. Xu, B. Xu, and B. He, An alternative subspace approach to EEG dipole source localization, *Physics in Medicine and Biology*, vol. 49, pp. 327–343, 2004.

chapter eight

EEG inverse problem IV
Bayesian techniques

Introduction

This chapter explains the basic mathematical formulation for the Bayesian framework, in general, and EEG source localization, in particular. The Bayesian framework is used to localize the brain sources within a probabilistic formulation that involves statistical terminology. Hence, the algorithm that was developed on the Bayesian framework is termed as multiple sparse priors (MSP) and is implemented in various publications. This chapter discusses in detail the MSP, its parameters, and its implementation for the EEG inverse problem. The topics covered are as follows: generalized Bayesian formulation, and then introduction to MSP with its fundamental mathematical derivations. After this, the essential parameters that are related to the performance of MSP are discussed in detail. The chapter discusses the modified MSP, which was first implemented by the author of this book. Hence, the method was termed as modified MSP, because the number of patches is subjected to change, and thus the method is compared with MSP and classical (minimum norm estimation [MNE], low-resolution brain electromagnetic tomography [LORETA], and beamformer) in terms of negative variational free energy (or simply free energy) and localization error. These terms are defined in detail in coming chapters.

8.1 Generalized Bayesian framework

To understand the Bayesian framework, we first need to understand the Bayes' theorem. Bayes' theorem defines the probability of an event, based on prior knowledge of conditions that might be related to the event. This theorem was proposed by Thomas Bayes (1701–1761), which provided an equation to allow new evidence for belief updates [1–5]. This theorem relies on the conditional probabilities for a set of events. The conditional probability for two events A and B is defined as "The conditional probability of B given A can be found by assuming that event A has occurred and, working under that assumption, calculating the probability that event B will occur." One of the ways to understand the Bayesian theorem is to know that we are dealing with sequential events, whereby new additional information is obtained for a subsequent event. Hence, this new information is used to revise the probability of the initial event. In this context, the terms *prior probability* and *posterior probability* are commonly used. Thus, before, explaining Bayesian theorem, some basic definitions are defined [6–10]:

- *Sample space*: The set of all possible outcomes of a statistical experiment is called the sample space and is represented by *S*.
- *Event*: The elements of sample space *S* are termed as events. In simple words, an event is a subset of a sample space *S*.
- *Intersection*: The intersection of two events A and B, denoted by the symbol $A \cap B$, is the event containing all elements that are common to A and B.
- *Mutually exclusive events*: Two events A and B are mutually exclusive or disjoint if $A \cap B = \phi$—that is, if A and B have no elements in common.
- *Prior probability*: A prior probability is an initial probability value originally obtained before any additional information is obtained.
- *Posterior probability*: A posterior probability is a probability value that has been revised by using additional information that is later obtained.

After presenting short definitions for the major terms involved in the formulation of Bayesian theorem, now the theorem is explained.

Let the *m* events $B_1, B_2, ..., B_m$ constitute a partition of the sample space *S*. That is, the B_i's are mutually exclusive such that

$$B_i \cap B_j = \phi \text{ for } i \neq j \tag{8.1}$$

and exhaustive:

$$S = B_1 \cup B_2 \cup ... \cup B_m \tag{8.2}$$

In addition, suppose the prior probability of the event B_i is positive— that is, $P(B_i) > 0$ for $i = 1, ..., m$. Now, if A is an event, then A can be written as the union of *m* mutually exclusive events, namely,

$$A = (A \cap B_1) \cup (A \cap B_2) \cup ... \cup (A \cap B_m) \tag{8.3}$$

Hence,

$$P(A) = P(A \cap B_1) + P(A \cap B_2) + \cdots + P(A \cap B_m) \tag{8.4}$$

Equation 8.4 can also be written as:

$$P(A) = \sum_{i=1}^{m} P(A \cap B_i) \tag{8.5}$$

$$\text{Or} \quad P(A) = \sum_{i=1}^{m} P(B_i) \times P(A \mid B_i) \tag{8.6}$$

Therefore, from these equations, the posterior probability of event B_k given event A is given by

$$P(B_k \mid A) = \frac{P(B_k \cap A)}{P(A)} \tag{8.7}$$

$$\text{Or} \quad P(B_k \mid A) = \frac{P(B_k) \times P(A \mid B_k)}{\sum_{i=0}^{m} P(B_i) \times P(A \mid B_i)} \tag{8.8}$$

This is termed as the Bayesian theorem and is extensively used in various fields for estimation purposes, which include the EEG inverse problem.

The Bayesian framework is preferred as it allows marginalizing noninteresting variables by integrating them out. Second, the stochastic sampling techniques such as Monte Carlo, simulated annealing genetic algorithms, and so on, are permissible under the Bayesian framework, and finally, it provides a posterior distribution of the solution (conditional expectation); in this aspect, the deterministic framework only allows for ranges of uncertainty. The prior probability of the source activity, $p(J)$, given by the previous knowledge of the brain behavior, is corrected for fitting the data using the likelihood $p(Y|J)$, allowing one to estimate the posterior source activity distribution using Bayes' theorem as [11,12]

$$p(J \mid Y) = \frac{p(J) \times p(Y \mid J)}{p(Y)} \tag{8.9}$$

Hence, the current density (J) is estimated using the expectation operator on posterior probability such that

$$\hat{J} = E[p(J \mid Y)] \tag{8.10}$$

The evidence $p(\mathbf{Y})$ is ignored due to its constant dataset values. Thus, Equation 8.10 becomes

$$p(\mathbf{J}\,|\,\mathbf{Y}) \propto p(\mathbf{Y}\,|\,\mathbf{J})p(\mathbf{J}) \qquad (8.11)$$

The noise that is associated with EEG measurements is assumed to be white Gaussian in nature. Thus, according to the generalized linear model as explained in previous chapters, the noise probability is given by

$$p(\in) = \mathcal{N}(\in;0,\Sigma_\in) \qquad (8.12)$$

where $\mathcal{N}(.)$ defines the multivariate Gaussian distribution and $\Sigma_\in$ is the variance associated with it with zero mean.

For multinormal distributions, in which multinormal distribution for a random variable $x \in R^{N\times1}$ with mean μ_x and covariance Σ_x is defined as [13,14]

$$p(x) = \mathcal{N}(x;\mu_x,\Sigma_x) = \frac{1}{(2\pi)^{N/2}|\Sigma_x|^{0.5}} \exp\left\{\frac{-1}{2}tr\left[(x-\mu_x)^T \sum_x^{-1}(x-\mu_x)\right]\right\} \qquad (8.13)$$

Here $tr(.)$ and $()^T$ are the trace and transpose operators, respectively. From Equation 8.13, it is evident that the solution is dependent on the covariance associated at the source or sensor level. Hence, the expected value and covariance for the given data can be found by taking the logarithmic operation of Equation 8.13 as:

$$\log p(x) \propto -\frac{1}{2}tr\left[(x-\mu_x)^T \sum_x^{-1}(x-\mu_x)\right] \qquad (8.14)$$

$$\propto -\frac{1}{2}tr\left[x^T\sum_x^{-1}x - x^T\sum_x^{-1}\mu_x - \mu_x^T\sum_x^{-1}x + \mu_x^T\sum_x^{-1}\mu_x\right]$$

Hence, the log of posterior probability distribution will be

$$\log p(\mathbf{J}\,|\,\mathbf{Y}) \propto \log p(\mathbf{Y}\,|\,\mathbf{J}) + \log p(\mathbf{J})$$

$$\propto -\frac{1}{2}tr\left[(\mathbf{Y}-\mathbf{LJ})^T\sum_\in^{-1}(\mathbf{Y}-\mathbf{LJ})\right] - \frac{1}{2}tr[(\mathbf{J}-\mu_J)^T\mathbf{Q}^{-1}(\mathbf{J}-\mu_J)] \qquad (8.15)$$

$$\propto -\frac{1}{2}tr\left[\mathbf{J}^T\left[\mathbf{L}^T\overset{-1}{\underset{\epsilon}{\sum}}\mathbf{L}+\mathbf{Q}^{-1}\right]\mathbf{J}-\mathbf{J}^T\left[\mathbf{L}^T\overset{-1}{\underset{\epsilon}{\sum}}\mathbf{Y}+\mathbf{Q}^{-1}\mu_J\right]\right]$$

$$-\left[\mathbf{Y}^T\overset{-1}{\underset{\epsilon}{\sum}}\mathbf{J}+\mu_J^T\right]+\left[\mathbf{Y}^T\overset{-1}{\underset{\epsilon}{\sum}}\mathbf{Y}+\mu_J^T\mathbf{Q}^{-1}\mu_J\right]$$

Hence, the covariance for posterior probability is given by

$$\mathrm{cov}[p(\mathbf{J}\,|\,\mathbf{Y})]=\sum_J=\left[L^T\overset{-1}{\underset{\epsilon}{\sum}}\mathbf{L}+\mathbf{Q}^{-1}\right]^{-1} \tag{8.16}$$

with expected value $\mathbf{E}[p(\mathbf{J}\mid\mathbf{Y})]$ as:

$$\mathbf{E}[p(\mathbf{J}\,|\,\mathbf{Y})]=\hat{\mathbf{J}}=\sum_J=\left[L^T\overset{-1}{\underset{\epsilon}{\sum}}\mathbf{Y}+\mathbf{Q}^{-1}\mu_J\right]$$

$$=\left[L^T\overset{-1}{\underset{\epsilon}{\sum}}\mathbf{L}+\mathbf{Q}^{-1}\right]^{-1}\left[L^T\overset{-1}{\underset{\epsilon}{\sum}}\mathbf{Y}+\mathbf{Q}^{-1}\mu_J\right] \tag{8.17}$$

Therefore, with some mathematical manipulations, the estimated current density for source reconstruction based on the Gaussian assumption with known values of source covariance ($\mathbf{Q}$) and prior sensor noise covariance matrix (Σ_e) is given by

$$\hat{\mathbf{J}}=\mathbf{Q}\mathbf{L}^T\left[\underset{\epsilon}{\sum}+\mathbf{L}\mathbf{Q}\mathbf{L}^T\right]^{-1}\mathbf{Y} \tag{8.18}$$

From Equation 8.18, it is evident that the accuracy of the three-dimensional (3D) solution for source estimation is highly dependent on the prior covariance matrix and prior noise covariance matrix, respectively. The solution in terms of optimization terminology can be written as:

$$\mathbf{J}_{estimated}=\arg\max_J\{p(\mathbf{J}\,|\,\mathbf{Y})\}=\arg\max_J\left\{\exp\left[-\frac{1}{2\sigma^2}\|\mathbf{Y}-\mathbf{L}\mathbf{J}\|_2^2-\lambda F(\mathbf{J})\right]\right\} \tag{8.19}$$

where $F(\mathbf{J})$ is a function that can be expressed as:

$$F(\mathbf{J})=\|\mathbf{W}\mathbf{J}\|_2^2 \tag{8.20}$$

Hence, from Equation 8.18, it is evident that proper selection of prior covariance matrices is necessary for the estimation of brain sources. The following discussion is presented for classical and MSP methods related to selection of prior covariance matrices.

8.2 Selection of prior covariance matrices

In this section, we initially discuss the prior sensor noise covariance matrix. In the case where there is no information about noise over sensors or their gain differences, the noise covariance is assumed to be of the following form:

$$\sum_{\in} = h_0 \mathbf{I}_{Nc} \tag{8.21}$$

where $\mathbf{I}_{Nc} \in \Re^{Nc \times Nc}$ is the identity matrix, and h_0 is the sensor noise variance.

In this formulation, the amount of noise variance is assumed to be uniform on all sensors. This is termed the *regularization parameter* [15] or *hyperparameter*. Thus, according to the literature [16], the prior information about noise can be employed through empty room recordings. In addition, some information for the estimation of empirical noise covariance can also be entered as an additional covariance component at the sensor level.

Continuing our discussion related to selection for prior covariance matrices, the source covariance matrix ($\mathbf{Q}$) can be derived through multiple constraints. According to the basic literature on source estimation [17,18], the simplest assumption for the sources says that all dipoles have the same prior variance and no covariance. This assumption is applied for classical MNE, and thus the prior source covariance is given by

$$\mathbf{Q} = \mathbf{I}_{N_d} \tag{8.22}$$

However, there exists another assumption in the literature, which states that the active sources vary smoothly within the solution space. This assumption was reported in the LORETA model [19,20]. For the smoothening purpose, Green's function is used [21]. This function is implemented using a Laplacian graph. Here, the faces and vertices are derived from structural magnetic resonance imaging. The Laplacian graph $G_L \in \Re^{N_d \times N_d}$ (here, N_d is the number of dipoles) is based on the adjacency matrix $A \in \Re^{N_d \times N_d}$, which is given by

$$G_{Lij} = \left\{ -\sum_{k=1}^{N_d} A_{ik}, \text{for } i = j \quad \text{and} \quad A_{ij}, \text{for } i \neq j \right. \tag{8.23}$$

Hence, Green's function $Q_G \in \Re^{N_d \times N_d}$ is defined as:

$$Q_G = e^{\sigma G_L} \tag{8.24}$$

Here, σ is a positive constant value that is used to determine the smoothness of the current distribution or spatial extent of the activated regions. The solution provided by LORETA is obtained using Green's function $Q = h_0 Q_G$, which shows that the LORETA uses a smooth prior covariance component unlike MNE, which uses an identity matrix.

Different improvements and modifications are suggested so far in the literature to normalize the noise or bias correcting mechanism. These modifications are introduced to obtain the solution with better resolution, as the resolution provided by LORETA and MNE is low. Hence, the latest framework is based on a Bayesian probabilistic model with modified source covariance matrix, as we see in the next section. This model is termed as an MSP and is explained in the following section.

8.3 Multiple sparse priors

The MSP model is based on the Bayesian framework as explained in the previous section. However, the selection of prior source covariance matrices is different when compared with the classical MNE and LORETA techniques. For MNE, the prior was an identity matrix, whereas for LORETA, it was designed as a fixed smoothing function that couples nearby sources. By contrast, MSP is based on a library of covariance components. Each of such covariance components corresponds to a different locally smooth focal region (which is termed as patch) of cortical surface. Hence, the generalized prior covariance matrix, taking into consideration the weighted sum of multiple prior components, that is, $C = \{C_1, C_2, \ldots, C_{N_q}\}$, is given by

$$Q = \sum_{i=1}^{N_q} h_i C_i \tag{8.25}$$

where $C_i \in \Re^{N_d \times N_d}$ is prior source covariance matrix and $h = \{h_1, h_2, \ldots, h_{N_q}\}$ is a set of hyperparameters; N_q is the number of patches or focal region. The set of hyperparameters is used to weigh the covariance components in such a way that regions having larger hyperparameters assign larger prior variances, and vice versa. It should be noted that these components may have different types of informative priors, which include different smoothing functions, medical knowledge, functional magnetic resonance imaging priors, and so on. Hence, the inversion model in general is dependent on the selection of prior components C, which ultimately refers to selection of prior assumptions [22]. Hence, the prior covariance matrix for MSP is generated by a linear mixture of covariance components from a library of priors.

To optimize the performance of MSP, the set of hyperparameters is subjected to optimization using a cost function. This cost function is derived using a probabilistic model employing a joint probability distribution with the inclusion of hyperparameters into a basic equation, that is, $p(\mathbf{Y}, \mathbf{J}, h)$. It is mentioned in the literature that the current density ($\mathbf{J}$) is dependent on the hyperparameters, and thus the model-based data are dependent on the current density. Hence, the following relationship is provided:

$$p(\mathbf{Y}, \mathbf{J}, h) = p(\mathbf{Y} \mid \mathbf{J})p(\mathbf{J} \mid h)p(h) \tag{8.26}$$

The aforementioned relation shows that prior distribution is dependent on hyperparameters as $p(\mathbf{J} \mid h)$. The generalized probability distribution for h is provided as:

$$p(h) \propto \prod_{i=1}^{N_q} e^{f_i(h_i)} \tag{8.27}$$

where each $f_i(\cdot)$ is a known unspecified function, which is mostly convex.

For a known probability distribution on h, the probability of hyperparameters is provided as:

$$p(\mathbf{J}) = \int p(\mathbf{J}, h)dh = \int p(\mathbf{J} \mid h)p(h)dh \tag{8.28}$$

The calculation of estimated values of hyperparameters is adopted in three different methods:

1. *Hyperparameter maximum a posteriori*: Here, h is assumed to be known and a solution is estimated for $\mathbf{J}$.
2. *Source maximum a posteriori*: Here, $\mathbf{J}$ is assumed to be known and the estimation is carried out for h.
3. *Variational Bayes' approximation*: Here, both $\mathbf{J}$ and h are based on the evidence ($\mathbf{Y}$), then by using the Laplace approximation, the factorization is carried out for joint probability such that $p(\mathbf{J}, h \mid \mathbf{Y}) \approx p(\mathbf{J} \mid \mathbf{Y})p(h \mid \mathbf{Y})$ and then the solution is provided.

These methods are covered in Wipf and Nagarajan [23] with extensive mathematical derivations and calculations.

According to the literature, if the hyperparameters are exclusively defined in terms of data, then their optimization can be defined as follows:

$$h' = \arg\max_{h} p(\mathbf{Y}, h) = \arg\max_{h} p(\mathbf{Y} \mid h)p(h) \tag{8.29}$$

Hence, the function defined above can be obtained by maximizing the following cost function:

$$\Theta(h) = \log p(\mathbf{Y} \mid h)p(h) \tag{8.30}$$

Assuming a multinormal distribution function for $p(\mathbf{Y} \mid h)$ and $p(h)$ as defined in Equation 8.27, we have

$$\Theta(h) = \log\left(\frac{1}{(2\pi)^{Nc/2}\sqrt{|\Sigma_Y|}}\right)\exp\left[-\frac{1}{2}tr\left(\mathbf{Y}^T\sum_Y^{-1}\mathbf{Y}\right)\right]\prod_{i=1}^{N_q}\exp[f_i(h_i)] \tag{8.31}$$

$$= -\frac{N_t}{2}tr\left(C_Y\sum_Y^{-1}\right) - \frac{N_t}{2}\log\left|\sum_Y\right| - \frac{N_cN_t}{2}\log(2\pi) + \sum_{i=1}^{N_q}f_i(h_i) \tag{8.32}$$

Here, $C_Y = (1/N_T)\mathbf{Y}\mathbf{Y}^T$ is the sample covariance.

According to various literature reviews [23], for the models that are based on the Gaussian approximation, the evidence $p(\mathbf{Y})$ is well approximated using the so-called negative variational free energy or simply free energy as the objective or cost function. In some research studies, the prior on the hyperparameters is assumed to be linear without prior expectations. However, Friston et al. provided an extended version of it by introducing a quadratic function with the inclusion of nonzero prior expectations in the basic formulation. The developed cost function was termed as free energy as it was derived from negative variational free energy maximization. Because this function is used for the quantification of localization results, the basic derivation is produced, as shown in the next section.

8.4 Derivation of free energy

The free energy is used as a cost function as mentioned above for models based on Gaussian distribution used for source estimation. The term *free energy* was derived from negative variational free energy maximization [24]. This cost function has an important role in defining the ability of any source estimation algorithm, so its basic derivation is discussed here.

First, define the log evidence as:

$$\log p(\mathbf{Y}) = \int q(h) \log p(\mathbf{Y}) dh \tag{8.33}$$

The joint probability distribution $p(\mathbf{Y}, \mathbf{J}, h)$ is defined as:

$$p(\mathbf{Y}, h) = \int p(\mathbf{Y}, \mathbf{J}, h) d\mathbf{J} = p(\mathbf{Y} \mid h) p(h) \tag{8.34}$$

$$
\begin{aligned}
\log p(y) &= \int q(h) \log \frac{p(\mathbf{Y}, h)}{p(h \mid \mathbf{Y})} dh \\
&= \int q(h) \log \frac{p(\mathbf{Y}, h) q(h)}{q(h) p(h \mid \mathbf{Y})} dh \\
&= \int q(h) \log \frac{p(\mathbf{Y}, h)}{q(h)} dh + \int q(h) \log \frac{q(h)}{p(h \mid \mathbf{Y})} dh \\
&= F + KL[q(h) \| p(h \mid \mathbf{Y})]
\end{aligned}
\tag{8.35}
$$

where KL is the Kullback–Leibler divergence [25] between the approximate $q(h)$ and posterior $p(h \mid \mathbf{Y})$, which is always positive and F is negative variational free energy. Given the condition, for $q(h)$ $p(h \mid \mathbf{Y})$, the KL becomes zero and thus the free energy is

$$F = \log p(\mathbf{Y}) \tag{8.36}$$

The purpose of any optimization algorithm is to maximize the value of F. Hence, further elaborating for derivation of free energy, it will become

$$
\begin{aligned}
F &= \int q(h) \log p(\mathbf{Y}, h) dh - \int q(h) \log q(h) dh \\
&= \Psi(\mathbf{Y}, h) + \mathbf{H}(h)
\end{aligned}
\tag{8.37}
$$

where $\Psi(\mathbf{Y}, h)$ is the expected energy and $\mathbf{H}(h)$ is entropy. The entropy is defined as:

$$\mathbf{H}(h) = -\int q(h) \log q(h) dh = \frac{N_p}{2} [1 + \log(2\pi)] + \frac{1}{2} \log \left| \sum_h \right| \tag{8.38}$$

For solving the entropy problem as energy cannot be integrated, the Laplace method approximation is applied. Thus, after performing the second-order Taylor series expansion for $\hat{U}(\mathbf{Y}, h)$, where $\hat{U}(\mathbf{Y}, h) = \log p(\mathbf{Y}, h)$, it will become

$$\hat{U}(\mathbf{Y}, h) = U(\mathbf{Y}, \hat{h}) + \frac{1}{2} tr\left[\left(h - \hat{h}\right)^T \mathbf{H}(h - \hat{h})\right] \tag{8.39}$$

The gradients are zero as it is performed at maximum; thus, the Hessian H is computed with

$$H_{i,j} = \frac{\partial^2 \hat{U}(\mathbf{Y}, h)}{\partial h_i \partial h_j} \tag{8.40}$$

Therefore, the approximated expected energy can be computed as:

$$
\begin{aligned}
\hat{\Psi} &= \int q(h) \hat{U}(\mathbf{Y}, h) dh \\
&= \int q(h) \left\{ \hat{U}(\mathbf{Y}, \hat{h}) + \frac{1}{2} tr\left[(h - \hat{h})^T \mathbf{H}\left(h - \hat{h}\right)\right]\right\} dh \\
&= \hat{U}(\mathbf{Y}, \hat{h}) \int q(h) dh + \frac{1}{2} \int q(h) tr\left[\left(h - \hat{h}\right)^T \mathbf{H}(h - \hat{h})\right] dh
\end{aligned}
\tag{8.41}
$$

By defining the posterior covariance of hyperparameters as:

$$\sum_h = \left(h - \hat{h}\right)^T \left(h - \hat{h}\right)$$

and inserting this into Equation 8.41, we get

$$
\begin{aligned}
\hat{\Psi} &= \hat{U}(\mathbf{Y}, \hat{h}) + \frac{1}{2} \int q(h) tr\left(\sum_h \mathbf{H}\right) dh \\
&= \hat{U}(\mathbf{Y}, \hat{h}) + \frac{1}{2} tr\left(\sum_h \mathbf{H}\right) \int q(h) dh \\
&= \hat{U}\left(\mathbf{Y}, \hat{h}\right) + \frac{1}{2} tr\left(\sum_h \mathbf{H}\right)
\end{aligned}
\tag{8.42}
$$

From the literature, the Hessian is replaced by the inverse of the posterior covariance $\mathbf{H} = -\Sigma_h^{-1}$:

$$\hat{\Psi} = \hat{U}\left(\mathbf{Y},\hat{h}\right) - \frac{1}{2}tr(I_{Np}) = \hat{U}\left(\mathbf{Y},\hat{h}\right) - \frac{N_p}{2} \tag{8.43}$$

Thus, the term on the left is the approximated evidence value, whereas the other defines the number of patches. This value can be approximated by

$$\hat{U}\left(\mathbf{Y},\hat{h}\right) = \log p\left(\mathbf{Y},\hat{h}\right) = \log p\left(\mathbf{Y}\,|\,\hat{h}\right)p_0\left(\hat{h}\right) \tag{8.44}$$

$$= \log \left(\begin{array}{c} \dfrac{1}{(2\pi)^{\frac{Nc}{2}}\sqrt{|\Sigma_Y|}}\exp\left[\dfrac{-1}{2}tr\left[Y^T\sum_Y^{-1}Y\right]\right] \\[4ex] \times\dfrac{1}{(2\pi)^{\frac{Np}{2}}\sqrt{|\Pi^{-1}|}}\exp\left\{-\dfrac{1}{2}tr\left[\left(\hat{h}-v\right)^T\Pi\left(\hat{h}-v\right)\right]\right\} \end{array} \right) \tag{8.45}$$

The term $p_0(h)$ is prior knowledge about hyperparameters, and $q(h)$ is Gaussian such that $p_0(h) = \mathcal{N}(h; v, \Pi^{-1})$ and $q(h) = \mathcal{N}(h; \hat{h}, \Sigma_h)$ with Π as prior precision vector of hyperparameters.

Therefore, after some mathematical manipulation by replacing H and $\hat{\Psi}$ as defined earlier, the free energy is written as:

$$F = -\frac{1}{2}tr\left[\mathbf{Y}^T\Sigma_Y^{-1}\mathbf{Y} + \left(\hat{h}-v\right)^T\Pi\left(\hat{h}-v\right)\right] - \frac{1}{2}\log|\Sigma_Y| - \frac{1}{2}\log|\Pi^{-1}| - \frac{Nc}{2}\log(2\pi)$$
$$-\frac{N_p}{2}\log(2\pi) - \frac{N_p}{2} + \frac{1}{2}\left\{N_p[1+\log(2\pi)] + \log|\Sigma_h|\right\} \tag{8.46}$$

By some more simplification and introducing the sample covariance matrix $C_Y = (1/N_t)YY^T$, the free energy will take the following form:

$$F = -\frac{N_t}{2}tr\left[C_Y\sum_Y^{-1}\right] - \frac{N_t}{2}\log\left|\sum_Y\right| - \frac{N_tN_n}{2}\log(2\pi)$$
$$-\frac{1}{2}tr\left[(\hat{h}-v)^T\Pi\,(\hat{h}-v)\right] + \frac{1}{2}\log\left|\sum_h\Pi\right| \tag{8.47}$$

The aforementioned cost function can be written in words as:

$$F = -[\text{Model Error}] - [\text{Size of model covariance}]$$
$$- [\text{Number of data samples}] - [\text{Error in hyperparameters}]$$
$$+ [\text{Error in covariance of hyperparameters}] \tag{8.48}$$

8.4.1 Accuracy and complexity

For simplicity, the free energy cost function is divided into two terms: accuracy and complexity. The accuracy term accounts for the difference between the data and estimated solution. This term is explained by model error, the size of the model-based covariance, and the number of data samples. However, the complexity is defined as the measurement for the optimization of hyperparameters—that is, the measure between the prior and posterior hyperparameter means and covariance. Hence, the accuracy and complexity are expressed as:

$$F(h) = \text{Accuracy}(h) - \text{Complexity}(h)$$

where

$$\text{Accuracy} = \frac{N_t}{2} tr\left(C_Y \sum_Y\right)^{-1} - \frac{N_t}{2} \log\left|\sum_Y\right| - \frac{N_c N_t}{2} \log(2\pi) \tag{8.49}$$

$$\text{Complexity} = -\sum_{i=1}^{N_p} f_i h_i \tag{8.50}$$

The complexity term is defined in different ways in different literature reviews; however, here it is included from López et al. [11]. From the equation, it is evident that there exists a trade-off between the accuracy and complexity. The accuracy is dependent on the number of covariance increased; however, at a certain point with an increase in covariance components, the accuracy increases but at the cost of high complexity. Therefore, a balance is maintained for the increase in the number of patches and thus covariance components to maintain a good trade-off between accuracy and complexity. This maintenance will bring a uniform change in free energy with optimization of a number of patches, and thus, the overall algorithm performance.

8.5 Optimization of the cost function

As discussed in previous sections, the accuracy of inversion methods is highly dependent on the selection of covariance matrices as well as priors. Different priors are assumed for various inversion methods as we discussed in the "Selection of Prior Covariance Matrices" section. However, by definition, the covariance components are defined as a set of "patches"; hence, it is necessary to evaluate which of them are relevant to the solution by increasing their h values [11]. Thus, to optimize the cost function, we need to optimize the h using any iterative algorithm such as expectation maximization (EM) [26]. Here the assumption of **J** as hidden data is carried out. The algorithm works as for the E-step, the hyperparameters are fixed, and hence the problem is solved using the following equations of the posterior covariance matrix of **J** such that

$$\mathrm{cov}\big[p(\mathbf{J}\,|\,\mathbf{Y})\big] = \sum_J = \left(L^T \sum_\epsilon^{-1} L + \mathbf{Q}^{-1}\right)^{-1} \tag{8.51}$$

$$\text{And } \hat{\mathbf{J}} = \mathbf{Q}L^T\left(\sum_\epsilon + LQL^T\right)^{-1} \mathbf{Y} \tag{8.52}$$

However, in the M-step, the hyperparameters are optimized with the gradient and Hessian of $\Theta(h)$, which is given as:

$$\Theta(h) = \log\left\{\frac{1}{(2\pi)^{\frac{Nc}{2}}\sqrt{\Sigma_Y}}\exp\left[\frac{-1}{2}tr\left(\mathbf{Y}^T\sum_Y^{-1}\mathbf{Y}\right)\right]\prod_{i=1}^{N_p}\exp(f_ih_i)\right\} \tag{8.53}$$

$$= -\frac{N_t}{2}tr\left(C_Y\sum_Y^{-1}\right) - \frac{N_t}{2}\log\left|\sum_Y\right| - \frac{N_cN_t}{2}\log(2\pi) + \sum_{i=1}^{N_p}f_ih_i \tag{8.54}$$

Hence, the first derivative of $\Theta(h) = F(h)$ as explained above with respect to hyperparameter (h_i) will be

$$\frac{\partial F(h)}{\partial h_i} = -\frac{N_t}{2}tr\big[P_i(C_Y - \Sigma_Y)\big] - \prod_{ii}(h_i - v_i) \tag{8.55}$$

with

$$P_i = \frac{\partial \Sigma_Y^{-1}}{\partial h_i} = h_i \sum_Y^{-1} LD_i L^T \sum_Y^{-1} \tag{8.56}$$

In addition, the Hessian is obtained by taking the derivative of the gradient:

$$\frac{\partial^2 F(h)}{\partial h_i \partial h_j} = -\frac{N_t}{2} tr(P_i D_i P_j D_i) - \prod_{ii} \tag{8.57}$$

From the aforementioned equations, it should be noted that the first term defines the accuracy, and the second term is related to the complexity. Hence, from the aforementioned relations it is clear that optimal values for hyperparameters will result in maximal values for free energy values, such that

$$\hat{h} = \arg\max_h F \tag{8.58}$$

Thus, the maximum of this function is located with a gradient ascent, which is dependent on the gradient and Hessian of the free energy as calculated above.

To increase the computational efficiency of the EM algorithm, a restricted maximum likelihood (ReML) was proposed [12]. This algorithm follows the steps as outlined:

- The model-based covariance $\Sigma_Y^{(K)}$ is calculated for the kth iteration. The hyperparameters are initialized with zero value provided that there are no informative hyperpriors.
- Evaluate the gradient as explained in the aforementioned equations for free energy for each hyperparameter. For the case where there is no hyperpriors, use $v = 0, \Pi \approx 0 I_{N_p}$.
- Calculate the curvature (Hessian) of the free energy as mentioned in the aforementioned equations for each hyperparameter.
- Update the hyperparameter such that

$$h_i^{(K)} = h_i^{(K-1)} + \Delta h_i \tag{8.59}$$

- In the above update, each hyperparameter variation, that is, Δh_i, is computed by Fisher scoring over the free energy variation such that

$$\Delta h_i = -\left[\frac{\partial^2 F(h)}{\partial h_i \partial h_j}\right]^{-1} \frac{\partial F(h)}{\partial h_i} \tag{8.60}$$

- Remove the hyperparameters with zero value and thus their corresponding covariance component. This is done because if a hyperparameter is zero, then the patch has no variance and it is removed.

- Finally, update free energy variation such that

$$\Delta F = \frac{\partial F}{\partial h} \Delta h \qquad (8.61)$$

Hence, the optimization process is topped here if the variation is less than a given tolerance; otherwise, start from Step 1.

8.5.1 Automatic relevance determination

In this way, ReML-based VL is used to optimize the hyperparameters for maximization of the free energy parameter for the MSP algorithm. However, to reduce the computational burden, automatic relevance determination (ARD) and greedy search (GS) algorithms are used for the optimization of sparse patterns. The implementation of ARD for MSP is defined in the following steps:

1. It uses ReML for the estimation of covariance components from the sample covariance matrix by taking into account the design matrix, eigenvalues matrix for basis functions, and number of samples. Hence, the outputs are estimated errors associated with each hyperparameter, ReML hyperparameters, free energy, accuracy, and complexity. The algorithm starts by defining the number of hyperparameters and then declaring uninformative hyperparameters. The design matrix is then composed by ortho-normalizing it. The scaling of the covariance matrix and the basis function calculation are performed next. After this step, the current estimates of covariance are computed using the EM algorithm. Thus, in the first step of the EM algorithm, the conditional variance is calculated followed by the next step, which estimates the hyperparameters with respect to free energy optimization. The final result is free energy calculation, which is subtraction of complexity from the accuracy term. The detailed outline is provided in pseudocode in the next section.
2. After ReML optimization, the spatial prior is designed using the number of patches (N_p) and optimized hyperparameters through ReML as discussed earlier.
3. Finally, the empirical priors are arranged and linearly multiplied by a modified leadfield to get optimized spatial components and thus inversion for source estimation.

In ARD, the singular value decomposition is used to generate new components for noise covariance and source covariance, respectively. Both of these components are considered in the measurement model such that the new model is given as:

$$\hat{\Sigma}_Y = \sum_{i=1}^{Nm+Np} h_i \hat{D}_i \tag{8.62}$$

where $\hat{D}$ is a new set of components and is defined as:

$$\hat{D}_{Nm+i} = L\left[\mathbf{Q}_{(.,i)}\mathbf{Q}_{(:,i)}^T\right]L^T \tag{8.63}$$

Here, $\mathbf{Q}$ is called the stacked matrix.

As explained earlier, the gradient and curvature (Hessian) of the free energy for each hyperparameter are computed separately, which increases the computational burden. However, with ARD, both the gradient and curvature are computed in a single step as follows.

- The computation of model-based sample covariance matrix $\hat{\Sigma}_Y^{(K)}$ for the kth iteration.
- Compute a stacked matrix $\mathbf{Q}^k$ such that its columns are formed from the main diagonal of each remaining component of a new set of components $\hat{D}$.
- Evaluate the gradient of the free energy as:

$$\frac{dF}{dh} = -\frac{1}{2}diag(h)\mathbf{Q}^T\left[\hat{\Sigma}_Y^{-1}C_Y\hat{\Sigma}_Y^{-1} - \hat{\Sigma}_Y^{-1}\right]\hat{Q} \tag{8.64}$$

- For the curvature of the free energy, use:

$$\frac{dF}{dh} = diag(h)\left[\left(\hat{Q}\hat{\Sigma}_Y^{-1}\hat{Q}\right)^T \cdot \hat{Q}\hat{\Sigma}_Y\hat{Q}\right]diag(h) \tag{8.65}$$

- Hence, update the hyperparameters:

$$h_i^{(k)} = h_i^{(k-1)} + \Delta h_i \tag{8.66}$$

where the change in each parameter Δh_i is computed by Fisher scoring over the free energy variation as:

$$\Delta h_i = -\left(\frac{\partial^2 F}{\partial h_i \partial h_j}\right)^{-1} \frac{\partial F}{\partial h_i} \tag{8.67}$$

Eliminate the hyperparameters that have values close to zero, and thus remove the matrices corresponding to them from the set of covariance components.

- Finally, update the free energy variation:

$$\Delta F = \frac{\partial F}{\partial h} \Delta h \tag{8.68}$$

It can be noted from the above algorithm that ARD works the same as ReML except that it computes the gradient and curvature of the free energy with respect to hyperparameters for all hyperparameters simultaneously. This reduces the computational time for simulation as compared with ReML. Further details for ARD are provided [27].

8.5.2 GS algorithm

GS works in a way by creating new hyperparameters with the partitioning of covariance component set [28–30]. ARD is considered more computationally efficient as it optimizes the hyperparameters with respect to free energy simultaneously through a gradient descent and removing the hyperparameters that are below threshold. This saves the time to calculate the free energy for each hyperparameters. GS does the single-to-many optimization of hyperparameters, which is initialized by including all covariance components. Thus, the iterative procedure is applied for removal of redundant components to take the solution.

The implementation of GS follows the protocol, which includes the following simple steps:

- The number of patches is taken equivalent to the length of source component as defined above.
- A matrix $\mathbf{Q}$ is defined whose rows are equivalent to a number of sources (Ns) and columns Np such that $\mathbf{Q} \in \mathfrak{R}^{Ns \times Np}$.
- After this, a subroutine is used for Bayesian optimization of a multivariate linear model with GS. This algorithm uses the multivariate Bayesian scheme to recognize the states of brain from neuroimages.

It resolves the ill-posed many-to-one mapping, from voxel values or data features to a target variable, using a parametric empirical or hierarchical Bayesian model. The inversion is carried out using standard variational techniques or EM to provide the model evidence and the conditional density of model parameters. The inputs are target vector, data feature matrix, confounds, patterns, noise covariance matrix, and number of GS iterations. However, the result is free energy, covariance partition indices, covariance hyperparameters, conditional expectation/variance of voxel weights, and empirical prior covariance for ordered and original space.

- The empirical priors are accumulated based on the calculations provided above. It depends on the output of the routine defined above and the matrix **Q** defined above.
- Finally, inversion is done for the subject using source space and modified leadfield as defined above.

The mathematical framework for GS is explained in the following section showing the derivations for each step of implementation.

The GS algorithm is based on the idea of applying ReML (or VL) with a reduced number of hyperparameters. Thus, the new model-based sample covariance matrix $\hat{\Sigma}_Y$ is formed, which is defined as:

$$\hat{\sum_Y} = h_0\Sigma_\epsilon + \sum_{i=1}^{Ng} h_i LQG_iQ^T L^T \tag{8.69}$$

where $G = \{G_1, G_2, \ldots, G_{Ng}\}$ with $G_i \in \Re^{Np \times Np}$, $\mathbf{Q} \in \Re^{Nd \times Np}$ is the single source covariance matrix with each column of it formed by a main diagonal of each covariance component matrix $\mathbf{D}_j$.

Thus, the sources are estimated within the space of covariance components as:

$$\hat{J}_Q = \left(\sum_{i=1}^{Ng} h_i G_i \right) Q^T L^T \hat{\sum_Y}^{-1} Y \tag{8.70}$$

The process follows the steps as outlined below for the GS optimization:

- For the kth iteration, solve the inverse problem with the hyperparameters obtained from ReML. The first iteration, $k = 1$ is computed with a single matrix G_1 and two hyperparameters—that is, h_0 for the sensor noise and h_1 for G_1.

- Eliminate matrices of G having low hyperparameter values. This procedure will reduce the search space, and thus computational burden.
- Generate a new G_t matrix with ones on diagonal values corresponding to most active components of $\hat{J}_Q^K$.
- Continue the process until the log evidence converges.

The successful optimization of GS suggests the recovery of neural activity with

$$\hat{J} = Q\hat{J}_Q \tag{8.71}$$

Hence, GS works upon the generation of new hyperparameter and new **G** matrix after iteration is performed.

8.6 Flowchart for implementation of MSP

By following the procedure as described earlier, MSP is applied to synthetic and real-time EEG data using ARD and GS optimization. Hence, the flowchart for MSP using ARD, and GS is produced to briefly explain the basic building blocks for MSP (Figure 8.1).

8.7 Variations in MSP

It can be observed from the earlier discussion that the localization capability is proportional to better selection of covariance matrices, prior information, and the number of patches used for inversion. Hence, the variation in basic framework of MSP can be carried out using a various number of patches to see the impact of it on free energy and localization error for different signal-to-noise ratio levels for synthetic and real EEG data, respectively. Hence, in this section, the impact of patches is discussed briefly.

The patches are defined for Bayesian-based models as the covariance components having an assumption that cortical currents have some local coherence within a distance of a few millimeters. Hence, this set of patches forms the search space for the inverse problem. For the MSP solution, the set of selected covariance components or patches is very important. With their increase, the search space for active sources is increased, thereby giving better results. In the absence of prior information—that is, size, shape, and location of neural current flow—the set of covariance components should ideally be composed of patches for all search space. This idea seems to be less feasible in terms of computational burden. However, a much smaller number of patches would be less feasible in terms of optimized solutions as the less targeted search space will reduce the free energy and thus the solution is less accurate.

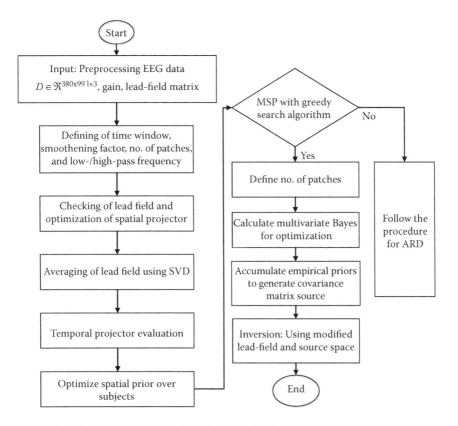

Figure 8.1 Flowchart for MSP (GS algorithm). ARD, Automatic relevance determination; MSP, multiple sparse priors; SVD, singular value decomposition.

There were attempts to generate the proper set of such patches for optimized solutions. In Harrison et al. [31], Green's function was defined to generate the set of patches. This function is based on a Laplacian graph and it forms a compact set of bell-shaped patches of finite cortical extent. The initial tests for such assumption have revealed that the neural sources that are far from the centers of defined patches using Green's function do not contribute in the source localization. Thus, the careful selection of patches has a major role to play in an optimized solution of the inverse problem.

This fact is exploited to develop modified MSP as a new method for increasing the number of patches. With the increase in the number of patches, the free energy (which defines accuracy and complexity for the system) is optimized (as seen in Chapter 9) and the localization error is reduced significantly. This shows that the number of patches will increase the system accuracy by optimizing the hyperparameters and thus the system solution. However, with an increase in the number of patches, the computational time is increased by a few seconds, which is minor as compared with optimization done in objective function (free energy) and reduction in localization error.

Summary

This chapter dealt with Bayesian framework-based inversion methods, which include MSP and its modified version. Hence, the discussion began with the introduction of Bayesian modeling in general. After this, Bayesian framework-based MSP was discussed, showing that the localization efficiency is dependent on covariance matrices. The cost function, that is, free energy, was discussed with mathematical derivations and theory. Furthermore, the optimization for cost function was discussed with ARD and GS algorithms. The impact of patches on localization was explored, and thus, a new method based on MSP, modified MSP, was discussed. Finally, the flow was defined for the implementation of MSP.

References

1. B. P. Carlin and T. A. Louis, *Bayes and Empirical Bayes Methods for Data Analysis*, Boca Raton, FL: CRC Press, 2000.
2. P. Smets, Belief functions: The disjunctive rule of combination and the generalized Bayesian theorem, *International Journal of Approximate Reasoning*, vol. 9(1), pp. 1–35, 1993.
3. G. Shafer, *A Mathematical Theory of Evidence*, Vol. 1, Princeton, NJ: Princeton University Press, 1976.
4. V. N. Vapnik and V. Vapnik, *Statistical Learning Theory*, Vol. 1. New York: Wiley, 1998.
5. G. D'Agostini, A multidimensional unfolding method based on Bayes' theorem, *Nuclear Instruments and Methods in Physics Research Section A: Accelerators, Spectrometers, Detectors and Associated Equipment*, vol. 362(2–3), pp. 487–498, 1995.
6. M. R. Spiegel et al., *Probability and Statistics*. New York: McGraw-Hill, 2009, ISBN: 978-0-07-154426-9.
7. A. Papoulis, *Probability & Statistics*, Vol. 2. Englewood Cliffs, NJ: Prentice Hall, 1990.
8. D. Dacunha-Castelle and D. Marie, *Probability and Statistics*, Vol. 2, Berlin, Germany: Springer Science & Business Media, 2012.
9. R. E. Walpole et al., *Probability and Statistics for Engineers and Scientists*, Vol. 5, New York: Macmillan, 1993.
10. J. S. Milton and J. C. Arnold, *Introduction to Probability and Statistics: Principles and Applications for Engineering and the Computing Sciences*, New York: McGraw-Hill, 2002.
11. J. López, V. Litvak, J. Espinosa, K. Friston, and G. R. Barnes, Algorithmic procedures for Bayesian MEG/EEG source reconstruction in SPM, *NeuroImage*, vol. 84, pp. 476–487, 2014.
12. K. Friston et al., Multiple sparse priors for the M/EEG inverse problem, *NeuroImage*, vol. 39, pp. 1104–1120, 2008.

13. N. R. Goodman, Statistical analysis based on a certain multivariate complex Gaussian distribution (an introduction), *The Annals of Mathematical Statistics*, vol. 34(1), pp. 152–177, 1963.

14. T. Koski, Multivariate Gaussian distribution, *Auxiliary Notes for Time Series Analysis SF2943*, Spring 2013.

15. G. Golub, Numerical methods for solving linear least squares problems, *Numerische Mathematik*, vol. 7, pp. 206–216, 1965.

16. J. D. López, J. J. Espinosa, and G. R. Barnes, Random location of multiple sparse priors for solving the MEG/EEG inverse problem, *Engineering in Medicine and Biology Society (EMBC), 2012 Annual International Conference of the IEEE*, San Diego, CA, IEEE, pp. 1534–1537, 2012.

17. M. Hämäläinen, R. Hari, R. J. Ilmoniemi, J. Knuutila, and O. V. Lounasmaa, Magnetoencephalography—Theory, instrumentation, and applications to noninvasive studies of the working human brain, *Reviews of Modern Physics*, vol. 65, p. 413, 1993.

18. S. Baillet, J. C. Mosher, and R. M. Leahy, Electromagnetic brain mapping, *Signal Processing Magazine, IEEE*, vol. 18, pp. 14–30, 2001.

19. R. D. Pascual-Marqui, Review of methods for solving the EEG inverse problem, *International Journal of Bioelectromagnetism*, vol. 1, pp. 75–86, 1999.

20. R. D. Pascual-Marqui et al., Low resolution brain electromagnetic tomography (LORETA) functional imaging in acute, neuroleptic-naive, first-episode, productive schizophrenia, *Psychiatry Research: Neuroimaging*, vol. 90, pp. 169–179, 1999.

21. K. Jan et al., A common formalism for the integral formulations of the forward EEG problem. *IEEE Transactions on Medical Imaging*, 24(1), pp. 12–28, 2005.

22. J. D. Lopez, G. R. Barnes, and J. J. Espinosa, Single MEG/EEG source reconstruction with multiple sparse priors and variable patches, *Dyna*, vol. 79, pp. 136–144, 2012.

23. D. Wipf and S. Nagarajan, A unified Bayesian framework for MEG/EEG source imaging, *NeuroImage*, vol. 44(3), pp. 947–966, 2009.

24. K. Friston, J. Mattout, N. Trujillo-Barreto, J. Ashburner, and W. Penny, Variational free energy and the Laplace approximation, *NeuroImage*, vol. 34, pp. 220–234, 2007.

25. J. M. Joyce, Kullback–Leibler divergence, in M. Lovric (ed.), *International Encyclopedia of Statistical Science*, Berlin, Germany: Springer, pp. 720–722, 2011.

26. T. K. Moon, The expectation-maximization algorithm, *IEEE Signal Processing Magazine*, vol. 13(6), pp. 47–60, 1996.

27. D. P. Wipf and S. S. Nagarajan, A new view of automatic relevance determination, *Advances in Neural Information Processing Systems*, pp. 1625–1632, 2008.

28. V. Litvak and K. Friston, Electromagnetic source reconstruction for group studies, *Neuroimage*, vol. 42(4), pp. 1490–1498, 2008.

29. P. Belardinelli et al., Source reconstruction accuracy of MEG and EEG Bayesian inversion approaches, *PLoS One*, vol. 7(12), p. e51985, 2012.

30. M. Haouari and J. S. Chaouachi, A probabilistic greedy search algorithm for combinatorial optimisation with application to the set covering problem, *Journal of the Operational Research Society*, vol. 53(7), pp. 792–799, 2002.

31. L. M. Harrison, W. Penny, J. Ashburner, N. Trujillo-Barreto, and K. Friston, Diffusion-based spatial priors for imaging, *NeuroImage*, vol. 38, pp. 677–695, 2007.

chapter nine

EEG inverse problem V
Results and comparison

Introduction

This chapter presents the results obtained by inversion of synthetic and EEG data from patients. In the first part of this chapter, the methodology is defined for the generation of synthetic EEG data at various signal-to-noise ratio (SNR) levels. Then, we move further to define the specifications of real-time EEG data and its sources. Furthermore, the results are provided for minimum norm estimation (MNE), low-resolution brain electromagnetic tomography (LORETA), beamformer, multiple sparse priors (MSP), and modified MSP in terms of free energy and localization error. It should be noted that the inversions are carried out for multiple runs to ensure the validity of the results statistically. The glass brain maps are also shown for some of the results. The advanced analysis for the developed algorithm is carried out using less electrodes for inversion. Hence, the results are summarized for that analysis as well.

9.1 Synthetic EEG data

This section is dedicated to defining the protocol to generate the synthetic EEG data at various SNR levels. For this, four levels are defined, which are 10, 5, 0, −5, and −20 dB, respectively. Hence, the protocol based on MATLAB programming is defined in the following section with various steps involved.

9.1.1 Protocol for synthetic data generation

- *Source generation:* This step defines the basics for the sources that are to be generated using the protocol. It defines the number of sources that are to be placed in the head model; their location, which can take any values; dipole amplitude; and the frequency of each dipole. It should be noted here that the number that defines the position of dipoles in the head model is critical and has a vital role in computation of localization error because it gives the three-dimensional (3D) coordinates for the exact location of dipoles. Thus, after estimation through inversion by any method, the Euclidean distance between estimated and actual is calculated to observe the localization error.

- *Waveform generation for each source:* This step corresponds to the generation of the waveform for each source. Thus, by providing amplitude, frequency, and phase of each dipole, the sinusoidal signal is plotted for each dipole. The resulting wave is sinusoidal because the signal generated from EEG should be varying in all times. However, because the data are synthetic, the signal is perfect sinusoidal. In a real case, the EEG signal is chaotic with a probabilistic nature.

- *Forward model generation:* This is the most critical and important step in the protocol for data generation. This step involves the calculation for gain matrix or leadfield matrix K. This matrix defines the relationship between the scalp measurements V and the current densities J. The leadfield matrix is constructed by imparting all the information related to channels and forward modality involved. The head geometry information is provided by a matrix named *vert*. This matrix has all the information related to coordinates for every point on the head—that is, if we put *vert(x)*, it gives us 3D coordinates for x point. This information is used for localization error as mentioned earlier. Hence, the leadfield matrix is computed. As a result of mesh generation, the triangular meshes are generated using the boundary element method as the head modeling scheme. Therefore, the distance matrix between each triangular mesh is calculated for smoothening of sources. Finally, data object D is generated, which contains information about channels, trials, data, etc.

- *Addition of noise:* After the pure signal generation, the next step is to introduce artificial noise into the data. This is done just to observe the change in the behavior of the inversion algorithms—that is, how they are localizing the sources in the presence of random noise. In practical terms, the noise is channel specific, but here it is introduced generally. Therefore, a Gaussian white noise is linearly added into the signal generated. The SNR can be altered to observe the effect of SNR on localization.

- *Data plots and object saving:* The last step for the protocol is plotting of the data. Hence, the data are plotted with and without noise. As explained earlier, the plots should be sinusoidal in nature. However, the generated data object D is saved, which will be used later for inversion purposes.

Thus, following this protocol, five synthetic EEG datasets are generated. These five datasets have the following specifications:

Dipole amplitude = 1×10^{-6} A
Dipole frequency = 20 Hz
Dipole positions = 2000, 5700
SNR = −5, 0, 5, 10 dB

However, to make the analysis stronger in terms of analysis, the algorithm is tested further on a low SNR value, that is, −20 dB. In addition, the dipole amplitude is reduced. Hence, the fifth dataset has the following features:

Dipole amplitude = 1×10^{-9} A
Dipole frequency = 20 Hz
Dipole positions = 2000, 5700
SNR = −20 dB

The location of the dipole is set according to the computed tomography fluoroscopy (CTF) head model and is estimated to be placed at both ends of scalp. However, they can vary and have no impact on the localization. For better visualization of the results, they are placed at these locations only.

After defining in detail the implementation protocol, now the flowchart is designed to define in a nutshell the implementation protocol for synthetic EEG data generation (Figure 9.1).

9.2 Real-time EEG data

This section defines the details of the real-time EEG data that are employed in this research work. In this research, multimodal face-evoked datasets are used for the inversion for all algorithms. This dataset is available online on the statistical parametric mapping (SPM) website which is kindly provided by Professor Rik Henson [1]. It is termed as multimodal because it has data from magnetometers (generating magnetoencephalography [MEG] data), gradiometers (also generating MEG data), and electrodes (generating EEG data), respectively. Many publications have followed this dataset for analyzing the localization for MEG and EEG modalities. The data were recorded by providing visual stimulus to healthy patients. The details of data are as follows:

- *Participants:* The participants were members of the MRC Cognition and Brain Sciences Unit participant panel. The multimodal data were compiled from a total of 19 participants (11 males and 8 females; aged 23–37 years, and of Caucasian origin). The study was approved by a Cambridge University Psychological Ethics Committee with written consent from participants.

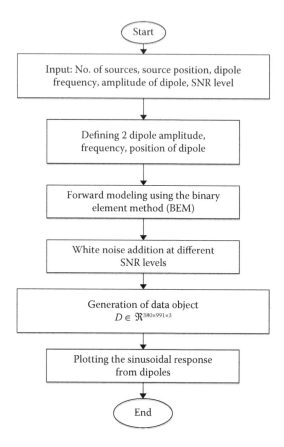

Figure 9.1 Flowchart for synthetic EEG data generation. SNR, Signal-to-noise ratio.

- *Stimuli:* The stimuli provided during this study were a series of faces, half of which were famous (known to participants) and half nonfamous (unknown to participants) personalities. A total of 300 faces with half male and half female were shown to the participants. Different variations were adapted for faces, which range from difference in hairstyle, expression change (both happy and neutral), and change in orientation with a majority of them in the full-frontal to three-fourth-view perspective. Besides this stimulus, another face dataset was created, which was called "scrambled faces" as it was generated by scrambling either famous faces or unfamiliar faces by taking two-dimensional (2D) Fourier transforms of them by permuting the phase information and then inverse transforming the back into image space. In this way, the power density spectrum of the image remained maintained. Thus, scrambled faces were subjected to cropping using a mask created by a combination of famous and unfamiliar faces. Hence, all of these stimuli were shown to healthy participants with the experimental design explained. Some samples from the dataset are also presented:

Unfamiliar faces

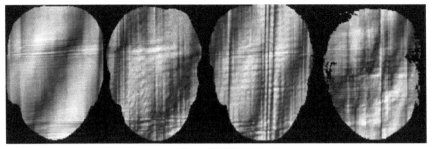

Scrambled faces

- *Experimental design*: The experimental setup was designed to provide the mentioned stimuli to participants, which was projected onto a screen that was held 1.3 m in front of them. Each image was presented against a black background, with a white fixation cross in the center. The trial was started when a fixation symbol changes from a circle to a cross. However, the mentioned stimulus was provided only after a certain time duration (here 500 ms), which lasts for 600 ms as shown in Figure 9.2. This fixation cross remained during the stimulus and for 700 ms afterward. The participants were told not to blink the eyes so as to reduce the electrooculography artifact as much as possible. In this way, the trials were designed for all stimuli. The stimulus-onset asynchrony was 4256 ms. Therefore, providing three types of faces (famous, unfamiliar, and scrambled) with different time intervals to various participants produces three trial types for each case. Thus, the data generated have three conditions with three trials, respectively. Therefore, the inversion can be checked for each participant and each trial and thus the variety of behavior can be analyzed for each algorithm.
- *M/EEG data acquisition*: The MEG and EEG data were captured in a slight magnetically shielded room using an Elekta Neuromag Vector View 306 Channel Meg system (Helsinki, Finland). The EEG was recorded using a 74-channel Easycap EEG cap. The standard 10%–10% electrode system was applied for readings. The head position indicator (HPI) coils were attached with the EEG cap to take simultaneous readings for MEG as well. The locations of EEG electrodes, the HPI coils, and 50–100 head points relative to the three anatomical

fiducials (the nasion and left and right preauricular points) were recorded using a 3D digitizer. The sampling rate was kept at 1100 Hz with a low-pass filter at 350 Hz and no high-pass filter. The reference electrode was placed at the nose, and the common ground electrode was placed at the left collarbone. The electrooculograms (vertical

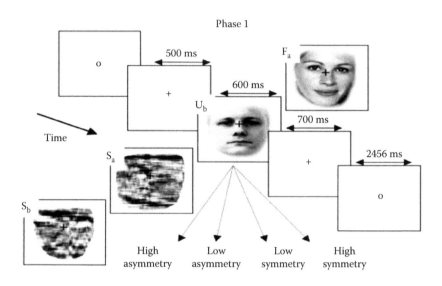

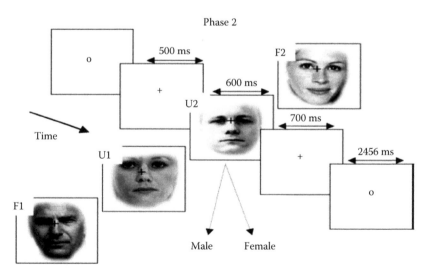

Figure 9.2 Experimental setup. (Adapted from *Statistical Parametric Mapping*. Available: http://www.fil.ion.ucl.ac.uk/spm/. Accessed on May 1, 2017.)

electrooculography and horizontal electrooculography) were measured by placing two sets of bipolar electrodes (Figure 9.2).

- *M/EEG preprocessing protocol*: The preprocessing stage has an important role in M/EEG data manipulation, because the data suffer from various artifacts and some channels should be deleted due to the artifact effect. The preprocessing protocol designed for this EEG data is defined in several steps, which include loading and converting of EEG data, epoching of data within a certain time frame, data downsampling to reduce the file size, artifact removal, filtering, taking grand mean for all participants, and finally generation of event-related potential (ERP) signal for all trials with three conditions as defined earlier with the saving of data into a ".mat" file. These steps are briefly discussed next.

  1. *Loading and conversion of data:* The raw M/EEG data are in biosemi format, which is not understandable by the SPM software on which the data are preprocessed. Therefore, it is essential to convert this format into an SPM-compatible format. Thus, ".bdf" is converted into ".mat" in the first step. The data file will now be opened in the SPM window and some basic information related to data loaded is shown. This information includes the sampling frequency, number of channels, and number of time samples. However, the other tabs show the name of channels, trials, inversion, and history of the data. The data plot can be seen using the "EEG" tab above the SPM screen.

  2. *Data epoching:* The epoching is done to cut little chunks of continuous data and save them in a single trial. For each stimulus onset, the epoched trial starts at some user-defined prestimulus time and ends at some poststimulus time. Thus, the resulting epoched data will also be baseline corrected, which implies that the mean of the prestimulus time is subtracted from the whole trial. For this dataset, a prestimulus time was defined at −100 and it ended at 600 ms. The epoching function in SPM performs baseline correction by default with baseline −200 to 0 ms. Thus, the epoched EEG data can be seen in the reviewing tool.

  3. *Data downsampling:* The downsampling is needed where the data are acquired at a higher sampling rate and thus occupy more space. Hence, if the sampling rate is higher, then downsampling is applied by providing a new sampling rate. It should be noted that the new sampling rate must be smaller than the old one. This is an additional preprocessing step that is required for making inferences about low-frequency components in the data.

  4. *Filtering:* The EEG data, whether it be continuous or epoched, is subjected to filtering. As defined in Section 2.1.1.2.1, the filtering

is needed to remove some unwanted low-frequency components (such as baseline shift signals having frequency <0.1 Hz) using high-pass filters, unwanted high-frequency components such as EMG artifact due to muscle activity where the frequency >30 Hz is removed using low-pass filters. In addition, in some cases where the data-acquisition system is not able to cancel out the effect of 50-Hz line noise, then a notch filter is used to mitigate this effect. Thus, here a "Butterworth" low-pass filter is applied with filter order of 5. The cutoff frequency is set at 32 Hz with the direction of filter as two pass, which means zero-phase forward and reverse filter. Thus, the filtered data are saved in an updated version of the data file.

5. *Artifact removal:* The artifacts are removed to clean the data and to make the observations more meaningful. Here the detection algorithm is applied on all channels with a threshold value of 10^{-4}. This will remove the channels having values greater than this threshold. Thus, the artifact caused due to eye blinking is removed by thresholding the data.

6. *Averaging:* The final step is generation of the ERP signal and for this averaging is done on the clean data. Hence, robust averaging is applied on data that will effectively suppress the artifacts without rejecting the trials or channels completely. The weights for the robust averaging are computed and saved automatically.

9.2.1 *Flowchart for real-time EEG data*

The flowchart representing the sequential flow for the preprocessing protocol is presented in Figure 9.3.

9.3 *Real-time EEG data results*

After presenting the protocol for the real-time EEG data, now the results and their corresponding discussions for the real-time EEG data are presented. Hence, the results are obtained for each inversion method, that is, MNE, LORETA, beamformer, MSP, and modified MSP. The quality of the methods is checked with the help of two parameters, that is, free energy that shows the accuracy and complexity as discussed in previous chapters. The localization error is not taken into consideration as the exact location of the dipole is unknown. Thus, the results are produced for each case. However, due to the large amount of graphs, the activation maps for all techniques are produced for Subject #01 only. For others, the tables are presented as the result by comparing the ability for various inversion techniques.

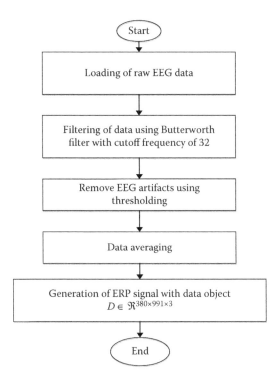

Figure 9.3 Flowchart for real-time EEG data. ERP, Event-related potential.

9.3.1 Subject #01: Results

The EEG data are generated using three trials ("famous," "unfamiliar," and "scrambled") at a sampling frequency of 1100 Hz by providing stimulus as defined in previous sections. The sample per trial rate is 991, which generates the object D, such that $D \in \Re^{380 \times 991 \times 3}$. The ERP generated from one channel is shown in Figure 9.4.

After the ERP signal is generated, inversions are applied as defined earlier. The results are presented in the following section.

9.3.2 Subject #01: Results for MSP, MNE, LORETA, beamformer, and modified MSP

The MSP is applied on Subject #01 of real-time EEG data. It shows the free energy level as −4153 with estimated time response at 3D coordinates of −12, −70, −9 mm, respectively. Thus, the computational time taken to compute the inversion for MSP was 8.1759 s. The resultant activation map is shown in Figure 9.5.

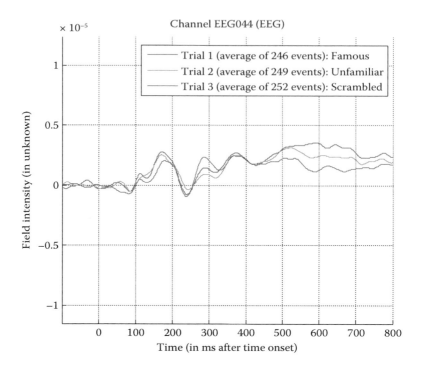

Figure 9.4 Event-related potential (ERP) signal for three-trial data.

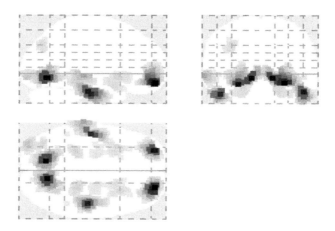

Figure 9.5 Multiple sparse priors (MSP) result for Subject #01.

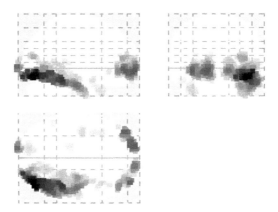

Figure 9.6 Low-resolution brain electromagnetic tomography (LORETA) result for Subject #01.

In a similar way, the LORETA results were obtained by applying the implementation protocol for LORETA, and the activation map is shown in Figure 9.6. It shows the estimated time response at the 3D coordinates 44, −80, −6 mm with free energy of −4932.0. Hence, the computational time required to perform this inversion was 6.5405 s.

The MNE is applied to check the behavior of the results, and it is observed that the estimated response time for MNE was at 3D coordinates of 48, −77, 0 mm, respectively, with free energy as −4944.2. However, the computational time was 5.7748 s. The resultant activation map is shown in Figure 9.7.

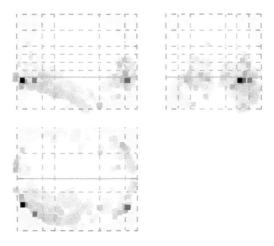

Figure 9.7 Minimum norm estimation (MNE) result for Subject #01.

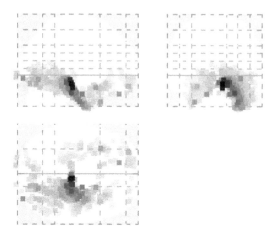

Figure 9.8 Activation map for beamformer Subject #01.

Finally, from the classical algorithms, the beamformer method was applied to Subject #01 data. The resultant activation map is shown in Figure 9.8, which shows the estimated response time at 3D coordinates of 15, −25, −12 mm, respectively. However, the free energy for this method is −4788.4. The computational time required for this method was 16.9116.

The results for the modified MSP are produced in the same pattern as they were produced for classical inversion techniques. Hence, the number of patches is increased and the result is produced for that particular patch's number. Thus, for the first time, the patches are increased to the level of 200, and the activation map is shown in Figure 9.9. The estimated time response was found at 3D coordinates of −54, −64, −15 mm, and the corresponding free energy is −4123.8. The computation time is 9.1537 s.

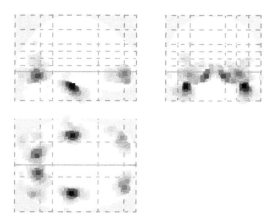

Figure 9.9 Modified multiple sparse priors (MSP) results for 200 patches.

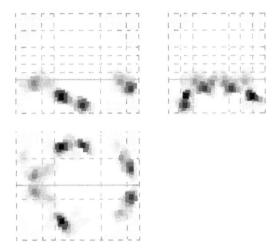

Figure 9.10 Modified multiple sparse priors (MSP) results for 300 patches.

In a similar way, the result was produced for patches at the level of 300. The free energy this time is −4111.0, while the estimated response time is observed at 3D coordinates of 58, −51, −20 mm. The computational time observed is 8.6611 s. The activation map is shown in Figure 9.10.

Again, the patches are increased up to 400, and the resulting activation map is shown in Figure 9.11. It shows the free energy as −4131.9 with response time at −40, −60, −20 mm. Thus, the time consumed for computation of this inversion is 9.0082 s.

In a similar way, the result was produced for patches at the level of 500. The free energy calculated is −4111.0, while the estimated response

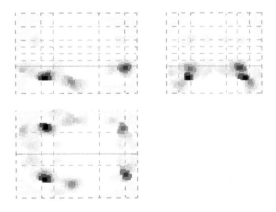

Figure 9.11 Modified multiple sparse priors (MSP) results for 400 patches.

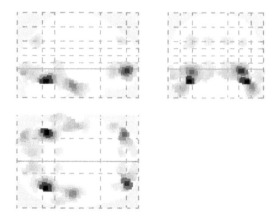

Figure 9.12 Modified multiple sparse priors (MSP) results for 500 patches.

time is observed at 3D coordinates of −40, −69, −19 mm. The computational time is 11.1227 s. The activation map is shown in Figure 9.12.

Continuing the procedure, the simulation was made for the patches level at 600 and the resulting activation map is shown in Figure 9.13. It shows the free energy as −4082.2 with a response time at −44, −74, −14 mm, respectively. Thus, the time consumed for computation is 11.4953 s.

In a similar way, the result was produced for 700 patches. The free energy calculated is −4104.5, while the estimated response time is observed at 3D coordinates of 36, −66, −12 mm. The computational time is 12.5123 s. The activation map is shown in Figure 9.14.

The simulations for patches up to 800 were carried out and the resulting activation map is shown in Figure 9.15. It shows the free energy as

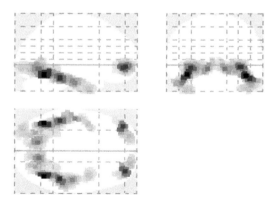

Figure 9.13 Modified multiple sparse priors (MSP) results for 600 patches.

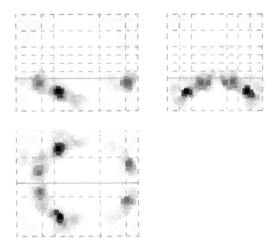

Figure 9.14 Modified multiple sparse priors (MSP) results for 700 patches.

−4113.5 with response time at 45, −81, −11 mm, respectively. Thus, the time consumed is 15.0699 s.

In a similar way, the result produced for 900 patches is optimized. The free energy calculated is −4115.0, while the estimated response time is observed at 3D coordinates of −54, −51, −21 mm. The computational time is 13.7438 s. The activation map is shown in Figure 9.16.

Following a similar pattern, the result was produced for 1000 patches. The free energy calculated is −4103.8, while the estimated response time is observed at 3D coordinates of −40, −65, −10 mm. The computational time is 14.7370 s. The activation map is shown in Figure 9.17.

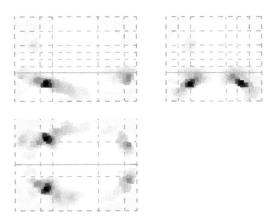

Figure 9.15 Modified multiple sparse priors (MSP) results for 800 patches.

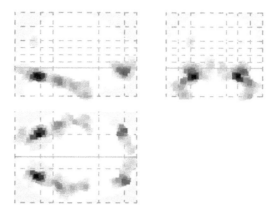

Figure 9.16 Modified multiple sparse priors (MSP) results for 900 patches.

In a similar way, the result was produced for the patches at a level of 1100. The free energy calculated is -4090.4, while the estimated response time is observed at 3D coordinates of 40, -65, -11 mm. The computational time is 16.4317 s. The activation map is shown in Figure 9.18.

Following a similar pattern, the result was produced for the patches at the level of 1200. The free energy calculated is -4101.7, while the estimated response time is observed at 3D coordinates of 37, -69, -15 mm. The computational time is 16.0366 s. The activation map is shown in Figure 9.19.

In a similar way, the result was produced for 1300 patches. The free energy calculated is -4122.0, while the estimated response time is observed at 3D coordinates of -22, -83, -16 mm. The computational time is 17.3896 s. The activation map is shown in Figure 9.20.

In a similar way, the result was produced for patches at the level of 1400. The free energy calculated is -4080.9, while the estimated response

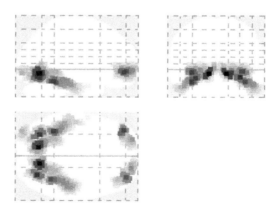

Figure 9.17 Modified multiple sparse priors (MSP) results for 1000 patches.

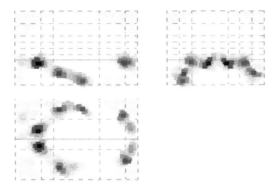

Figure 9.18 Modified multiple sparse priors (MSP) results for 1100 patches.

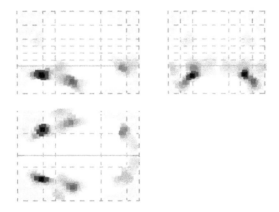

Figure 9.19 Modified multiple sparse priors (MSP) results for 1200 patches.

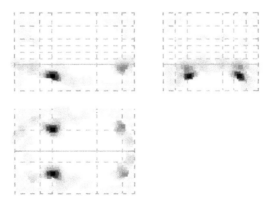

Figure 9.20 Modified multiple sparse priors (MSP) results for 1300 patches.

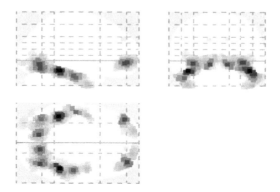

Figure 9.21 Modified multiple sparse priors (MSP) results for 1400 patches.

time is observed at 3D coordinates of 37, −69, −15 mm. The computational time is 19.0568 s. The activation map is shown in Figure 9.21.

Following a similar pattern, the result was produced for patches optimized at the level of 1500. The free energy calculated is −4100.2, while the estimated response time is observed at 3D coordinates of −39, 52, 0 mm. The computational time is 20.9768 s. The activation map is shown in Figure 9.22.

In a similar way, the result was produced for 1600 patches. The free energy calculated is −4100.2, while the estimated response time is observed at 3D coordinates of 55, −53, −23 mm. The computational time is 22.3799 s. The activation map is shown in Figure 9.23.

Following a similar pattern, the result was produced for patches at the level of 1700. The free energy calculated is −4093.0, while the estimated

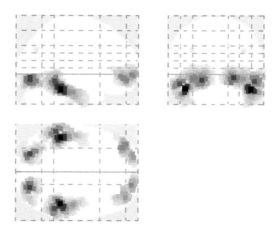

Figure 9.22 Modified multiple sparse priors (MSP) results for 1500 patches.

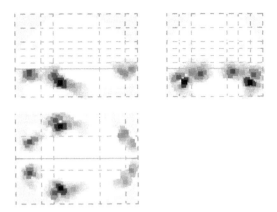

Figure 9.23 Modified multiple sparse priors (MSP) results for 1600 patches.

response time is observed at 3D coordinates of -39, 52, 1 mm. The computational time is 24.9116 s. The activation map is shown in Figure 9.24.

In a similar way, the result was produced for 1800 patches. The free energy calculated is -4098.0, while the estimated response time is observed at 3D coordinates of 44, -80, -6 mm. The computational time is 23.8903 s. The activation map is shown in Figure 9.25.

Following a similar pattern, the result was produced for 1900 patches. The free energy calculated is -4116.1, while the estimated response time is observed at 3D coordinates of 20, -55, -10 mm. The computational time is 24.3126 s. The activation map is shown in Figure 9.26.

Similarly, the result was produced for 2000 patches. The free energy calculated is -4111.6, while the estimated response time is observed at 3D coordinates of -33, 45, -5 mm. The computational time is 25.5685 s. The activation map is shown in Figure 9.27.

Finally, the result was produced for patches at the level of 2100. The free energy calculated is -4157.6, while the estimated response time is

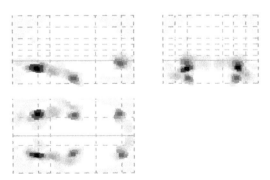

Figure 9.24 Modified multiple sparse priors (MSP) results for 1700 patches.

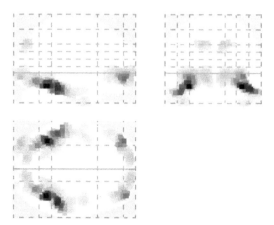

Figure 9.25 Modified multiple sparse priors (MSP) results for 1800 patches.

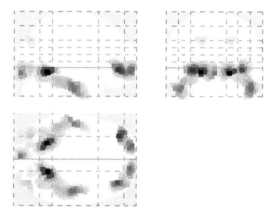

Figure 9.26 Modified multiple sparse priors (MSP) results for 1900 patches.

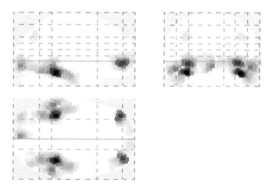

Figure 9.27 Modified multiple sparse priors (MSP) results for 2000 patches.

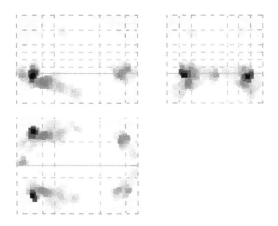

Figure 9.28 Modified multiple sparse priors (MSP) results for 2100 patches.

observed at 3D coordinates of −46, −79, −4 mm. The computational time is 29.5846 s. The activation map is shown in Figure 9.28.

The results obtained above are summarized in Table 9.1. This table provides the methods with their respective parameters of free energy and the computational time.

In a similar way as explained for Subject #01, the ERP data are generated using three trials ("famous," "unfamiliar," and "scrambled") at a sampling frequency of 1100 Hz by providing stimulus as defined in Chapter 3. The sample per trial rate is 991, which generates the object D, such that $D \in \Re^{380 \times 991 \times 3}$. The ERP generated from one channel is shown in Figure 9.29.

Thus, the inversion methods are applied one by one and the resulting parameters of free energy and computational complexity are computed for each. These results are summarized in Table 9.2.

In a similar way, the ERP data are generated using three trials ("famous," "unfamiliar," and "scrambled") at a sampling frequency of 1100 Hz by providing stimulus as defined previously for Subject #03. The sample per trial rate is 991, which generates the object D, such that $D \in \Re^{380 \times 991 \times 3}$. The ERP generated from one channel is shown in Figure 9.30.

Hence, after the data object is generated, it is loaded into MATLAB and the inversions are carried out for all methods. The results thus obtained are summarized in Table 9.3.

For Subject #04, the ERP data are generated using three trials ("famous," "unfamiliar," and "scrambled") at a sampling frequency of 1100 Hz by providing stimulus as explained before. The sample per trial rate is 991, which generates the object D, such that $D \in \Re^{380 \times 991 \times 3}$. The ERP generated from one channel is shown in Figure 9.31.

Table 9.1 Comparison between Various Methods for Subject #01

S. No.	Inversion Method	Free Energy	Computational Complexity (s)
01.	MSP	−4153.0	8.1759
02.	LORETA	−4932.0	6.5425
03.	MNE	−4944.2	5.7748
04.	Beamformer	−4788.4	16.9116
05.	Modified MSP (200 patches)	−4123.8	9.1500
06.	Modified MSP (300 patches)	−4111.0	8.6600
07.	Modified MSP (400 patches)	−4131.9	9.0082
08.	Modified MSP (500 patches)	−4111.0	11.1227
09.	Modified MSP (600 patches)	−4082.2	11.4953
10.	Modified MSP (700 patches)	−4104.5	12.5123
11.	Modified MSP (800 patches)	−4113.5	15.0699
12.	Modified MSP (900 patches)	−4115.0	13.7438
13.	Modified MSP (1000 patches)	−4103.8	14.7370
14.	Modified MSP (1100 patches)	−4090.4	15.2315
15.	Modified MSP (1200 patches)	−4101.7	16.0366
16.	Modified MSP (1300 patches)	−4122.0	17.3890
17.	Modified MSP (1400 patches)	−4080.9	19.0568
18.	Modified MSP (1500 patches)	−4100.2	20.9768
19.	Modified MSP (1600 patches)	−4100.0	22.3705
20.	Modified MSP (1700 patches)	−4093.0	24.9110
21.	Modified MSP (1800 patches)	−4098.0	23.8903
22.	Modified MSP (1900 patches)	−4116.1	24.3126
23.	Modified MSP (2000 patches)	−4111.6	25.5685
24.	Modified MSP (2100 patches)	−4157.6	29.5846

LORETA, Low-resolution brain electromagnetic tomography; MNE, minimum norm estimation; MSP, multiple sparse priors.

Hence, the localization algorithms are applied on these data and the results are compared in terms of free energy and the computational time. The results are summarized in Table 9.4.

Continuing the discussion for Subject #05, the ERP data are generated using three trials ("famous," "unfamiliar," and "scrambled") at a sampling frequency of 1100 Hz by providing stimulus. The sample per trial rate is 991, which generates the object D, such that $D \in \Re^{380 \times 991 \times 3}$. The ERP generated from one channel is shown in Figure 9.32.

Hence, all the estimation algorithms are applied on these data and the results are compared in terms of free energy and the computational time. These results are summarized in Table 9.5.

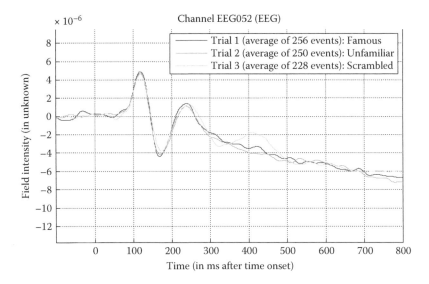

Figure 9.29 Event-related potential (ERP) signal for three-trial data.

Again, following the same procedure as done for previous subjects, the ERP data are generated using three trials ("famous," "unfamiliar," and "scrambled") at a sampling frequency of 1100 Hz by providing stimulus. The sample per trial rate is 991, which generates the object D, such that $D \in \mathfrak{R}^{380 \times 991 \times 3}$. The ERP generated from one channel is shown in Figure 9.33.

After this, the inversion methods are applied one by one for this dataset and the changes in free energy and computational time were noted. The results thus obtained are summarized in Table 9.6.

The ERP data for Subject #07 were generated by following the same preprocessing steps as mentioned earlier. Thus, the trials are the same—that is, three trials ("famous," "unfamiliar," and "scrambled")—at a sampling frequency of 1100 Hz by providing the same stimulus as described earlier. The sample per trial rate is 991, which generates the object D, such that $D \in \mathfrak{R}^{380 \times 991 \times 3}$. The ERP generated from one channel is shown in Figure 9.34.

Table 9.7 presents a comparison between the various algorithms.

The ERP data for Subject #08 are generated using the same preprocessing steps as mentioned earlier. Thus, the stimulus is also the same for these subject data. Thus, the object $D \in \mathfrak{R}^{380 \times 991 \times 3}$ is generated at a sampling frequency of 1100 Hz. The sample per trial rate is 991. The ERP signal from one channel looks like that shown in Figure 9.35.

The inversions are applied on these data to check the varying behavior for two parameters—that is, free energy and the computational complexity. Table 9.8 summarizes the results obtained from inversions.

Table 9.2 Comparison between Various Methods for Subject #02

S. No.	Inversion Method	Free Energy	Computational Complexity (s)
01.	MSP	−4468.9	11.5104
02.	LORETA	−5133.4	6.9738
03.	MNE	−5149.2	6.1803
04.	Beamformer	−5026.0	18.1674
05.	Modified MSP (200 patches)	−4419.0	9.1722
06.	Modified MSP (300 patches)	−4436.0	10.6721
07.	Modified MSP (400 patches)	−4432.9	11.1127
08.	Modified MSP (500 patches)	−4419.0	11.2633
09.	Modified MSP (600 patches)	−4435.7	11.6277
10.	Modified MSP (700 patches)	−4429.8	16.3000
11.	Modified MSP (800 patches)	−4421.4	12.5938
12.	Modified MSP (900 patches)	−4430.2	15.2861
13.	Modified MSP (1000 patches)	−4441.9	16.1284
14.	Modified MSP (1100 patches)	−4417.2	15.5539
15.	Modified MSP (1200 patches)	−4416.1	16.1284
16.	Modified MSP (1300 patches)	−4429.3	17.9703
17.	Modified MSP (1400 patches)	−4445.3	19.4105
18.	Modified MSP (1500 patches)	−4411.2	20.3104
19.	Modified MSP (1600 patches)	−4427.8	23.4258
20.	Modified MSP (1700 patches)	−4432.6	23.1612
21.	Modified MSP (1800 patches)	−4424.1	26.4683
22.	Modified MSP (1900 patches)	−4444.2	25.7404
23.	Modified MSP (2000 patches)	−4429.2	26.1042
24.	Modified MSP (2100 patches)	−4456.8	35.2719

LORETA, Low-resolution brain electromagnetic tomography; MNE, minimum norm estimation; MSP, multiple sparse priors.

Going ahead for the results from Subject #09, the same preprocessing steps were followed to generate object D such that $D \in \Re^{380 \times 991 \times 3}$. Thus, the sampling frequency was 1100 Hz with 991 samples/trials and a total of three ("famous," "unfamiliar," and "scrambled") trials. The ERP signal from one of the channels is shown in Figure 9.36.

In a similar way, for the results from Subject #10, the same preprocessing steps were followed to generate object D such that $D \in \Re^{380 \times 991 \times 3}$. Thus, the sampling frequency was 1100 Hz with 991 samples/trials and a total of three ("famous," "unfamiliar," and "scrambled") trials. The ERP signal from one of the channels is shown in Figure 9.37.

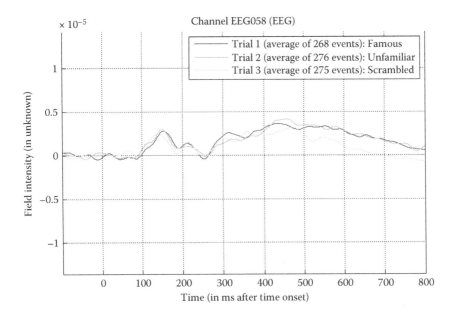

Figure 9.30 Event-related potential (ERP) signal for three-trial data.

9.4 Detailed discussion of the results from real-time EEG data

This section summarizes the results from various subjects' inversions. It can be seen from the results that with the increase in the number of patches, the free energy is improved significantly. In addition, the localization capability of increased number of patches MSP, which is termed as modified MSP, is much higher as compared with the classical MNE, LORETA, and beamformer methods. This fact is asserted for all subjects. Starting the discussion for Subject #01, it is clear from Table 9.1 that the proposed modified MSP has the highest accuracy as it has the highest free energy values for all instances as compared with the classical algorithms. The classical algorithms like LORETA, MNE, and beamformer have low free energy values, which are −4932.0, −4944.2, and −4788.4, respectively. These values are very low as compared with the best case for the modified MSP—that is, at 1100 patches where the value is −4090.4. Thus, by comparing the Bayesian algorithm-based MSP with the developed modified MSP, it can be seen that the modified MSP has higher free energy values, as MSP has −4153.0, which is lower than most of the values produced by modified MSP in the table. On average, the difference between free energy as provided by modified MSP and MSP is 20–30 units with maximum

Table 9.3 Comparison between Various Methods for Subject #03

S. No.	Inversion Method	Free Energy	Computational Complexity (s)
01.	MSP	−3841.7	9.6195
02.	LORETA	−4562.2	6.8775
03.	MNE	−4573.9	6.0568
04.	Beamformer	−4476.8	17.8555
05.	Modified MSP (200 patches)	−3813.7	10.9949
06.	Modified MSP (300 patches)	−3807.0	10.8359
07.	Modified MSP (400 patches)	−3810.4	10.1041
08.	Modified MSP (500 patches)	−3806.4	11.7722
09.	Modified MSP (600 patches)	−3785.4	12.2307
10.	Modified MSP (700 patches)	−3799.1	14.6864
11.	Modified MSP (800 patches)	−3798.9	16.0064
12.	Modified MSP (900 patches)	−3800.4	15.7040
13.	Modified MSP (1000 patches)	−3812.5	16.4080
14.	Modified MSP (1100 patches)	−3809.4	17.2159
15.	Modified MSP (1200 patches)	−3798.7	18.0226
16.	Modified MSP (1300 patches)	−3810.5	18.3721
17.	Modified MSP (1400 patches)	−3784.0	22.5793
18.	Modified MSP (1500 patches)	−3800.9	23.2031
19.	Modified MSP (1600 patches)	−3803.3	25.602
20.	Modified MSP (1700 patches)	−3806.6	27.9040
21.	Modified MSP (1800 patches)	−3812.2	26.6482
22.	Modified MSP (1900 patches)	−3815.8	30.7255
23.	Modified MSP (2000 patches)	−3822	26.690
24.	Modified MSP (2100 patches)	−3816.0	32.544

LORETA, Low-resolution brain electromagnetic tomography; MNE, minimum norm estimation; MSP, multiple sparse priors.

difference of 73 units. However, the difference between modified MSP and classical algorithms is 500–700 units, which is very large in terms of accuracy. It shows a clear supremacy of modified MSP over all other inversions. Continuing the discussion for the computational complexity parameter, the classical algorithms have less computational complexity. It is because of their simple algorithm steps unlike the Bayesian framework-based MSP and proposed modified MSP methods, respectively. Therefore, comparing the time required for computation of the inversion, the MSP and modified MSP have time variation from 15 to 20 s from classical algorithms. In the real-time case, this time variation is not very large

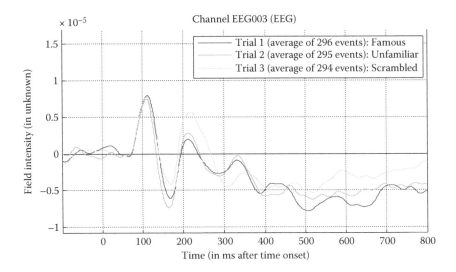

Figure 9.31 Event-related potential (ERP) signal for three-trial data.

as it is less than 1 min. It can be noted from the table that higher values of patches require larger times to compute the inversion. Thus, the time increases linearly with the increase in the patches, number. Therefore, it is concluded that the modified MSP performs well in terms of accuracy and complexity.

Again, for Subject #02, a similar trend was observed. Table 9.2 shows that the modified MSP has the highest accuracy as it has the highest free energy values for all runs as compared with the classical algorithms. The classical algorithms like LORETA, MNE, and beamformer have low free energy values, which are −5133.4, −5149.4, and −5026.2, respectively. These values are very low as compared with the best case for the modified MSP—that is, at 1500 patches where the value is −4411.2. However, comparing MSP with the developed modified MSP, it has higher free energy values as MSP has −4468.9, which is lower than most of the values produced by the modified MSP in the table. On average, the difference between free energy as provided by the modified MSP and MSP is 10–55 units with maximum difference of 57 units. However, the difference between modified MSP and classical algorithms is 600–800 units, which is very large in terms of accuracy. Thus, these results show a better performance of modified MSP over other algorithms. The other parameter covered for this discussion is the computational complexity parameter. The classical algorithms have lower computational complexity as compared with MSP and its modified version, respectively. It is because of their simple algorithm steps unlike the Bayesian framework-based

Table 9.4 Comparison between Various Methods for Subject #04

S. No.	Inversion Method	Free Energy	Computational Complexity (s)
01.	MSP	−4083.5	9.1216
02.	LORETA	−4447.1	6.8896
03.	MNE	−4451.7	6.1837
04.	Beamformer	−4334.9	18.0416
05.	Modified MSP (200 patches)	−4085.1	10.8151
06.	Modified MSP (300 patches)	−4076.9	10.8359
07.	Modified MSP (400 patches)	−4059.3	12.4224
08.	Modified MSP (500 patches)	−4058.2	14.4992
09.	Modified MSP (600 patches)	−4074.8	13.8783
10.	Modified MSP (700 patches)	−4059.5	16.5909
11.	Modified MSP (800 patches)	−4071.9	17.8092
12.	Modified MSP (900 patches)	−4086.6	19.4808
13.	Modified MSP (1000 patches)	−4098.9	17.8556
14.	Modified MSP (1100 patches)	−4075.7	19.8707
15.	Modified MSP (1200 patches)	−4078.7	21.4698
16.	Modified MSP (1300 patches)	−4063.4	21.5059
17.	Modified MSP (1400 patches)	−4104.6	25.7402
18.	Modified MSP (1500 patches)	−4053.6	23.0597
19.	Modified MSP (1600 patches)	−4077.8	26.3120
20.	Modified MSP (1700 patches)	−4076.7	27.5174
21.	Modified MSP (1800 patches)	−4077.1	27.5667
22.	Modified MSP (1900 patches)	−4094.3	33.2270
23.	Modified MSP (2000 patches)	−4074.5	26.7637
24.	Modified MSP (2100 patches)	−4117.9	33.7671

LORETA, Low-resolution brain electromagnetic tomography; MNE, minimum norm estimation; MSP, multiple sparse priors.

MSP and proposed modified MSP methods, respectively. Therefore, by comparing the time required for computation of the inversion, the MSP and modified MSP have been found to have time variation from 10 to 30 s from classical algorithms. However, in the real-time case, this time variation does not matter much as it is less than 1 min. As with the previous observation, it is observed that the higher values of patches require a larger time to compute the inversion. Thus, the time increases linearly with the increase in the patches number.

Continuing the discussion for Subject #03, it can be seen from Table 9.3 that the modified MSP has the highest accuracy as it has the highest free energy values for all iterations as compared with the classical algorithms.

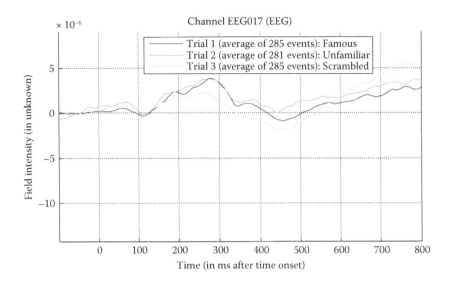

Figure 9.32 Event-related potential (ERP) signal for three-trial data.

The classical algorithms like LORETA, MNE, and beamformer have low free energy values, which are $-4462, -4573.9$, and -4476.9, respectively. These values are very low as compared with the best case for modified MSP—that is, for 1400 patches where the value is -3784.0. However, comparing MSP with modified MSP, even modified MSP has higher free energy values as MSP has -3841.7, which is lower than all of the values produced by modified MSP in the table. On average, the difference between free energy as provided by the modified MSP and MSP is 25–55 units with maximum difference of 55 units. However, the difference between modified MSP and classical algorithms is 700 units approximately, which is very large in terms of accuracy. Thus, it shows a better performance of modified MSP over other algorithms. Going through the computational complexity parameter, the classical algorithms have lower computational complexity as compared with MSP and modified MSP, respectively. The lower computational complexity is because of their simpler implementation strategy as compared with MSP and modified MSP, respectively. Therefore, comparing the time required for computation of the inversion, the MSP and modified MSP have time variation from 6 to 25 s from classical algorithms. However, in the real-time case, this time variation does not matter much as it is less than 1 min. Thus, similar to the observation made for Subject #01 and Subject #02, it is observed that the higher values of patches require larger times to compute the inversion. Thus, the time increases linearly with the increase in the patches' number.

For Subject #04, it is observed from Table 9.4 that the modified MSP has the highest accuracy as it has the highest free energy values for all

Table 9.5 Comparison between Various Methods for Subject #05

S. No.	Inversion Method	Free Energy	Computational Complexity (s)
01.	MSP	−5256.0	13.9259
02.	LORETA	−5751.0	6.5400
03.	MNE	−5761.5	5.9728
04.	Beamformer	−5650.1	16.9877
05.	Modified MSP (200 patches)	−5130.1	9.4341
06.	Modified MSP (300 patches)	−5129.0	8.5154
07.	Modified MSP (400 patches)	−5107.3	11.2750
08.	Modified MSP (500 patches)	−5125.8	10.5581
09.	Modified MSP (600 patches)	−5121.4	10.8454
10.	Modified MSP (700 patches)	−5094.7	14.5025
11.	Modified MSP (800 patches)	−5116.1	13.6740
12.	Modified MSP (900 patches)	−5128.5	13.6997
13.	Modified MSP (1000 patches)	−5147.9	15.4155
14.	Modified MSP (1100 patches)	−5114.2	17.8599
15.	Modified MSP (1200 patches)	−5124.9	17.3424
16.	Modified MSP (1300 patches)	−5121.4	19.4777
17.	Modified MSP (1400 patches)	−5144.4	19.6183
18.	Modified MSP (1500 patches)	−5116.3	19.4759
19.	Modified MSP (1600 patches)	−5119.0	21.3468
20.	Modified MSP (1700 patches)	−5127.6	22.3236
21.	Modified MSP (1800 patches)	−5124.1	23.8016
22.	Modified MSP (1900 patches)	−5178.7	26.1328
23.	Modified MSP (2000 patches)	−5137.7	25.0946
24.	Modified MSP (2100 patches)	−5132.0	26.5234

LORETA, Low-resolution brain electromagnetic tomography; MNE, minimum norm estimation; MSP, multiple sparse priors.

instances as compared with the classical algorithms. The classical algorithms like LORETA, MNE, and beamformer have low free energy values, which are −4447.1, −4451.7, and −4334.9, respectively. These values are very low as compared with the best case for the modified MSP—that is, with 1500 patches where the value is −4053.6. However, comparing MSP with modified MSP reveals that modified MSP has higher free energy for most instances than MSP except in two instances (1400 and 2100 patches). Thus, it is concluded that modified MSP performs well in terms of higher free energy. On average, the difference between free energy as provided by modified MSP and MSP is 5–30 units with maximum difference of 30 units approximately. However, the difference between modified MSP

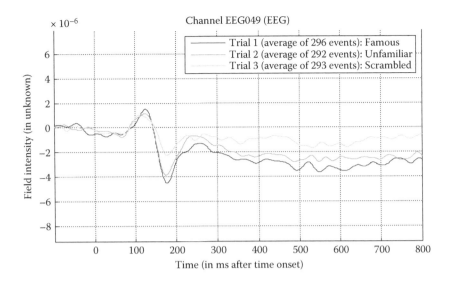

Figure 9.33 Event-related potential (ERP) signal for three-trial data.

and classical algorithms is 300–400 units, which is very large in terms of accuracy. As described earlier for the computational complexity discussion, the same holds true for these subject data; that is, the classical algorithms have less computational complexity. It is because of their simple algorithm steps unlike the Bayesian framework-based MSP and proposed modified MSP methods, respectively. Thus, by comparing the time required for computation of the inversion, the MSP and modified MSP have time variation from 15 to 25 s from classical algorithms. In the real-time case, this time variation is not very large as it is less than 1 min. It can be noted from the table that higher values of patches require larger times to compute the inversion.

Looking into the observations for Subject #05, it is clear that modified MSP has the highest accuracy as it has the highest free energy values for all instances as compared with the classical algorithms. The classical algorithms like LORETA, MNE, and beamformer have low free energy values, which are −5751.0, −5761.5, and −56.50, respectively. These values are very low as compared with the best case for the modified MSP, that is, at for 700 patches where the value is −5094.7. However, comparing MSP with modified MSP, modified MSP has higher free energy values for all instances. On average, the difference between free energy as provided by modified MSP and MSP is 109–162 units with maximum difference of 162 units. This is a very large difference as compared with previous subjects' data. However, the difference between the modified MSP and classical algorithms is 500–600 units, which is a huge difference

Table 9.6 Comparison between Various Methods for Subject #06

S. No.	Inversion Method	Free Energy	Computational Complexity (s)
01.	MSP	−3186.5	9.8886
02.	LORETA	−3731.7	6.6486
03.	MNE	−3737.0	5.8419
04.	Beamformer	−3654.1	17.3045
05.	Modified MSP (200 patches)	−3114.6	10.1843
06.	Modified MSP (300 patches)	−3128.1	14.1146
07.	Modified MSP (400 patches)	−3113.3	14.4601
08.	Modified MSP (500 patches)	−3084.0	17.5878
09.	Modified MSP (600 patches)	−3116.2	14.4282
10.	Modified MSP (700 patches)	−3127.2	18.4780
11.	Modified MSP (800 patches)	−3112.2	18.8163
12.	Modified MSP (900 patches)	−3132.4	19.9138
13.	Modified MSP (1000 patches)	−3113.7	20.5759
14.	Modified MSP (1100 patches)	−3128.8	24.0164
15.	Modified MSP (1200 patches)	−3098.3	28.1641
16.	Modified MSP (1300 patches)	−3127.6	22.9892
17.	Modified MSP (1400 patches)	−3117.9	26.8767
18.	Modified MSP (1500 patches)	−3114.2	27.8829
19.	Modified MSP (1600 patches)	−3109.4	29.5337
20.	Modified MSP (1700 patches)	−3105.4	27.3671
21.	Modified MSP (1800 patches)	−3134.0	31.8288
22.	Modified MSP (1900 patches)	−3117.9	31.2156
23.	Modified MSP (2000 patches)	−3178.6	33.6586
24.	Modified MSP (2100 patches)	−3112.0	34.7375

LORETA, Low-resolution brain electromagnetic tomography; MNE, minimum norm estimation; MSP, multiple sparse priors.

in terms of accuracy. The other parameter covered for this discussion is the computational complexity parameter. The classical algorithms have lower computational complexity as compared with MSP and modified MSP, respectively. It is because of their simple algorithm steps unlike the Bayesian framework-based MSP and proposed modified MSP methods, respectively. Thus, by comparing the time required for computation of the inversion, the MSP and modified MSP have time variation from 10 to 20 s from classical algorithms. However, in the real-time case, this time variation does not matter much as it is less than 1 min. As with the previous observation, it is observed that the higher values of patches require larger

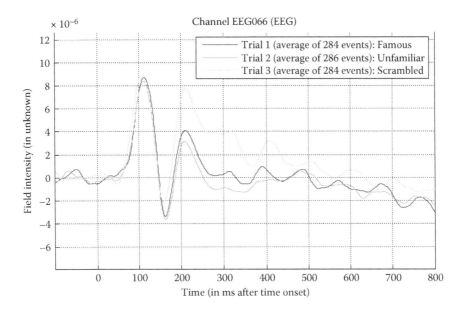

Figure 9.34 Event-related potential (ERP) signal for three-trial data.

times to compute the inversion. Thus, the time increases linearly with the increase in the patches number.

Continuing the discussion for Subject #06, it is observed from Table 9.6 that modified MSP has the highest accuracy as it has the highest free energy values for all instances as compared with the classical algorithms like other subjects. The classical algorithms like LORETA, MNE, and beamformer have low free energy values, which are −3731.7, −3731.0, and −3654.1, respectively. These values are very low as compared with the best case for the modified MSP—that is, at 500 patches where the value is −3084.0. However, upon comparing MSP with modified MSP, it is revealed that modified MSP has higher free energy values as MSP has −3186.5, which is lower than all of the values produced by modified MSP in the table. On average, the difference between free energy as provided by modified MSP and MSP is 10–102 units with maximum difference of 102 units. However, the difference between the modified MSP and classical algorithms is 600 units approximately, which is very large in terms of accuracy. Discussing for the computational complexity parameter, the classical algorithms have lower computational complexity as compared with MSP and modified MSP, respectively. Therefore, by comparing the time required for computation of the inversion, the MSP and modified MSP have time variation from 5 to 24 s from classical algorithms. However, in the real-time case, this time variation does not matter much as it is less than 1 min. Thus, similar to the

Table 9.7 Comparison between Various Methods for Subject #07

S. No.	Inversion Method	Free Energy	Computational Complexity (s)
01.	MSP	−5601.4	11.4035
02.	LORETA	−6541.4	7.8884
03.	MNE	−6545.8	6.7126
04.	Beamformer	−6384.7	19.0912
05.	Modified MSP (200 patches)	−5584.4	10.8891
06.	Modified MSP (300 patches)	−5562.4	8.5319
07.	Modified MSP (400 patches)	−5568.5	11.1008
08.	Modified MSP (500 patches)	−5572.0	10.7926
09.	Modified MSP (600 patches)	−5564.6	10.8399
10.	Modified MSP (700 patches)	−5574.7	13.2618
11.	Modified MSP (800 patches)	−5570.6	13.6594
12.	Modified MSP (900 patches)	−5566.5	14.6083
13.	Modified MSP (1000 patches)	−5589.6	15.1679
14.	Modified MSP (1100 patches)	−5582.2	16.5799
15.	Modified MSP (1200 patches)	−5571.9	16.5092
16.	Modified MSP (1300 patches)	−5549.3	18.0831
17.	Modified MSP (1400 patches)	−5537.9	20.5557
18.	Modified MSP (1500 patches)	−5564.8	23.0831
19.	Modified MSP (1600 patches)	−5575.3	23.6301
20.	Modified MSP (1700 patches)	−5583.7	23.4884
21.	Modified MSP (1800 patches)	−5561.0	24.4916
22.	Modified MSP (1900 patches)	−5590.4	25.9831
23.	Modified MSP (2000 patches)	−5524.0	26.0046
24.	Modified MSP (2100 patches)	−5582.1	27.2433

LORETA, Low-resolution brain electromagnetic tomography; MNE, minimum norm estimation; MSP, multiple sparse priors.

observation made for previous subjects, it is observed that the higher values of patches require a larger time to compute the inversion.

Furthermore, going through the results for Subject #07, it is evident that the proposed modified MSP has the highest accuracy as it has the highest free energy values for all instances as compared with the classical algorithms. The classical algorithms like LORETA, MNE, and beamformer have low free energy values of −6541.4, −6545.8, and −6384.7, respectively. These values are very low as compared to the best case for the modified MSP—that is, at 2000 patches where the value is −5524.0. However, comparing the MSP with the modified MSP, the modified MSP has higher free energy values as the MSP has −5601.4, which is lower than all of the

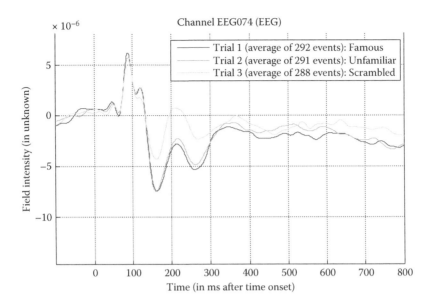

Figure 9.35 Event-related potential (ERP) signal for three-trial data.

values produced by the modified MSP in the table. On average, the difference between free energy as provided by the modified MSP and MSP is 10–55 units with maximum difference of 77 units. However, the difference between the modified MSP and classical algorithms is 1000 units roughly, which is very large in terms of accuracy. Hence, the results are showing better performance of modified MSP over other algorithms. Going through the computational complexity parameter, the classical algorithms have lower computational complexity as compared to MSP and modified MSP, respectively. Thus, comparing the time required for computation of the inversion, the MSP and modified MSP have a time variation from 5 to 22 s from classical algorithms. So, similar to observations made for previous subjects' data, it is observed that the higher values of patches require a larger amount of time to compute the inversion.

The discussion for Subject #08 is presented here. The results show that the proposed modified MSP has the highest accuracy as it has the highest free energy values for all instances as compared to the classical algorithms. The classical algorithms like LORETA, MNE, and beamformer have low free energy values, which are −4174, −4181.2, and −4111.4, respectively. These values are very low as compared with the best case for the modified MSP—that is, at 2200 patches where the value is −3763. However, as compared with MSP, the modified MSP has higher free energy values. On average, the difference between free energy as provided by modified MSP and MSP is 10–43 units with maximum difference of 43 units. However,

Table 9.8 Comparison between Various Methods for Subject #08

S. No.	Inversion Method	Free Energy	Computational Complexity (s)
01.	MSP	−3806.6	12.0733
02.	LORETA	−4174.0	6.6911
03.	MNE	−4181.2	5.7881
04.	Beamformer	−4111.4	17.3291
05.	Modified MSP (200 patches)	−3806.0	9.3817
06.	Modified MSP (300 patches)	−3795.0	9.7417
07.	Modified MSP (400 patches)	−3794.0	11.5431
08.	Modified MSP (500 patches)	−3774.5	12.4394
09.	Modified MSP (600 patches)	−3779.5	12.8615
10.	Modified MSP (700 patches)	−3778.0	14.9379
11.	Modified MSP (800 patches)	−3781.3	16.2660
12.	Modified MSP (900 patches)	−3766.4	17.6538
13.	Modified MSP (1000 patches)	−3779.1	17.6051
14.	Modified MSP (1100 patches)	−3781.4	17.8627
15.	Modified MSP (1200 patches)	−3779.3	18.9281
16.	Modified MSP (1300 patches)	−3768.5	19.8506
17.	Modified MSP (1400 patches)	−3780.2	20.0857
18.	Modified MSP (1500 patches)	−3763.9	23.2291
19.	Modified MSP (1600 patches)	−3772.0	25.6411
20.	Modified MSP (1700 patches)	−3772.8	26.0834
21.	Modified MSP (1800 patches)	−3769.2	28.5319
22.	Modified MSP (1900 patches)	−3778.2	29.0269
23.	Modified MSP (2000 patches)	−3777.0	32.7505
24.	Modified MSP (2100 patches)	−3763.0	29.3912

LORETA, Low-resolution brain electromagnetic tomography; MNE, minimum norm estimation; MSP, multiple sparse priors.

the difference between the modified MSP and classical algorithms is 400 units, which is a huge difference in terms of accuracy. The other parameter covered for this discussion is the computational complexity parameter. The classical algorithms have lower computational complexity as compared with MSP and modified MSP, respectively. Hence, upon comparison it is revealed that the time required for computation of the inversion, the MSP and modified MSP have time variation from 10 to 25 s from classical algorithms. As with the previous observations, it is observed that the higher values of patches require a larger time to compute the inversion.

This section provides a detailed discussion of the results from real-time EEG data for Subject #09. Table 9.9 shows that the modified MSP has

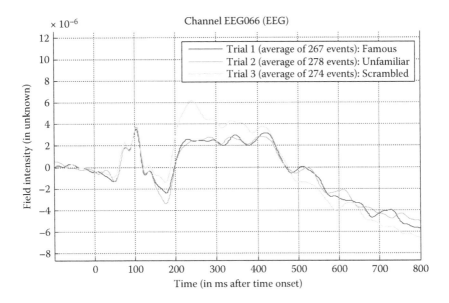

Figure 9.36 Event-related potential (ERP) signal for three-trial data.

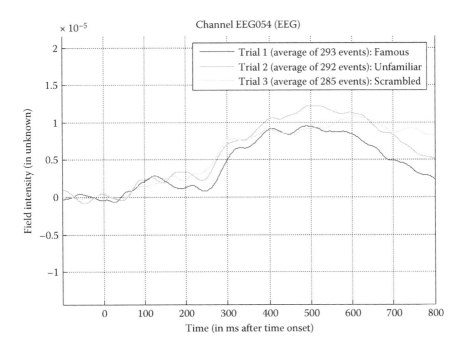

Figure 9.37 Event-related potential (ERP) signal for three-trial data.

Table 9.9 Comparison between Various Methods for Subject #09

S. No.	Inversion Method	Free Energy	Computational Complexity (s)
01.	MSP	−4227.8	10.8480
02.	LORETA	−5265.0	7.4408
03.	MNE	−5270.3	6.4486
04.	Beamformer	−5144.4	18.8001
05.	Modified MSP (200 patches)	−4237.2	11.5452
06.	Modified MSP (300 patches)	−4234.5	12.6740
07.	Modified MSP (400 patches)	−4235.8	12.7180
08.	Modified MSP (500 patches)	−4233.9	15.7014
09.	Modified MSP (600 patches)	−4245.5	13.9701
10.	Modified MSP (700 patches)	−4215.6	17.2036
11.	Modified MSP (800 patches)	−4228.0	19.0490
12.	Modified MSP (900 patches)	−4231.4	20.1022
13.	Modified MSP (1000 patches)	−4224.9	19.0376
14.	Modified MSP (1100 patches)	−4228.7	21.9076
15.	Modified MSP (1200 patches)	−4224.9	21.4364
16.	Modified MSP (1300 patches)	−4204.8	21.8648
17.	Modified MSP (1400 patches)	−4225.5	26.6334
18.	Modified MSP (1500 patches)	−4204.3	22.6529
19.	Modified MSP (1600 patches)	−4230.7	26.0704
20.	Modified MSP (1700 patches)	−4225.3	25.9840
21.	Modified MSP (1800 patches)	−4235.8	28.3206
22.	Modified MSP (1900 patches)	−4210.4	27.9864
23.	Modified MSP (2000 patches)	−4221.6	27.9379
24.	Modified MSP (2100 patches)	−4208.1	29.3816

LORETA, Low-resolution brain electromagnetic tomography; MNE, minimum norm estimation; MSP, multiple sparse priors.

the highest accuracy as it has the highest free energy values as compared with the classical algorithms. The classical algorithms like LORETA, MNE, and beamformer have low free energy values, which are −5265.0, −5270.3, and −5144.4, respectively. These values are very low as compared with the best case for the modified MSP—that is, at 1500 patches where the value is −4204.3. However, upon comparison with MSP, modified MSP has higher free energy values for most of the instances in the table. The maximum difference between the MSP and modified MSP is 24 units. However, the difference between modified MSP and classical algorithms is 1000 units approximately, which is a huge difference in terms of accuracy. As with the previous discussion, the other parameter covered for this discussion

Table 9.10 Comparison between Various Methods for Subject #10

S. No.	Inversion Method	Free Energy	Computational Complexity (s)
01.	MSP	−3629.7	12.2263
02.	LORETA	−4319.6	7.4637
03.	MNE	−4321.4	6.4193
04.	Beamformer	−4265.7	18.4311
05.	Modified MSP (200 patches)	−3597.5	12.8999
06.	Modified MSP (300 patches)	−3592.9	11.2727
07.	Modified MSP (400 patches)	−3586.6	14.8648
08.	Modified MSP (500 patches)	−3587.2	12.3692
09.	Modified MSP (600 patches)	−3591.7	15.3279
10.	Modified MSP (700 patches)	−3586.8	14.8175
11.	Modified MSP (800 patches)	−3582.6	15.4595
12.	Modified MSP (900 patches)	−3593.9	18.4270
13.	Modified MSP (1000 patches)	−3591.1	17.2485
14.	Modified MSP (1100 patches)	−3589.3	17.8092
15.	Modified MSP (1200 patches)	−3582.3	18.3805
16.	Modified MSP (1300 patches)	−3580.7	20.9566
17.	Modified MSP (1400 patches)	−3609.2	27.7304
18.	Modified MSP (1500 patches)	−3599.0	23.3074
19.	Modified MSP (1600 patches)	−3588.2	20.6070
20.	Modified MSP (1700 patches)	−3608.4	23.8988
21.	Modified MSP (1800 patches)	−3582.9	25.1544
22.	Modified MSP (1900 patches)	−3598.8	28.5768
23.	Modified MSP (2000 patches)	−3585.4	25.8065
24.	Modified MSP (2100 patches)	−3594.8	25.8262

LORETA, Low-resolution brain electromagnetic tomography; MNE, minimum norm estimation; MSP, multiple sparse priors.

is the computational complexity parameter. The classical algorithms have lower computational complexity as compared with the MSP and modified MSP, respectively. Hence, by comparing the time required for inversion, the MSP and modified MSP have a time variation from 10 to 25 s from classical algorithms.

In the end, a discussion is provided for Subject #10. From Table 9.10, it is evident that the modified MSP has the highest accuracy as it has the highest free energy values as compared with the classical algorithms. The classical algorithms like LORETA, MNE, and beamformer have low free energy values, which are −4319.6, −4321.4, and −4265.7, respectively. These values are very low as compared with the best case for the modified

MSP—that is, at 1300 patches where the value is -3580.7. In addition, compared with MSP, modified MSP has higher free energy values due to the higher number of patches. On average, the difference between free energy as provided by the modified MSP and MSP is 20–50 units with maximum difference of 50 units. However, the difference between the modified MSP and classical algorithms is on average 700 units approximately, which is very large in terms of accuracy. Going through the computational complexity parameter, the classical algorithms have lower computational complexity as compared with the MSP and modified MSP, respectively, as was observed for all other subjects.

9.5 Results for synthetic data

In this section, the results for synthetically generated EEG data are discussed. The synthetic EEG data are generated by following the protocol defined in Section 9.1. The source position is placed at 2000 and 5700 in source space. These positions are placed arbitrarily and can be changed in code. These locations are further used to compute the localization error using the Euclidean distance formula defined in the next section.

9.5.1 Localization error

The brain source localization techniques are compared in terms of accuracy, which is defined by localization error. The localization error is the difference between the simulated and estimated locations, respectively. Thus, the smallest value defines the highest accuracy, and vice versa. This parameter is, however, valid only for synthetically generated EEG data as we are aware of the exact location. Hence, mathematically, this parameter is written as:

$$\text{Localization Error} = \| S_{\text{true}} - S_{\text{estimated}} \| \tag{9.1}$$

where S_{true} is the exact location of dipole in x, y, z coordinates and $S_{\text{estimated}}$ is the estimated position in x, y, z coordinates using any of the algorithms defined earlier. Hence, this parameter provides a 3D comparison between various algorithms for synthetic EEG data.

9.5.2 Synthetic data results for SNR = 5 dB

The MSP, LORETA, MNE, and beamformer algorithms are applied to synthetic EEG data generated at an SNR level of 5 dB. The algorithms are run multiple times, and thus the data are averaged for free energy and the localization error at dipole 2000 and dipole 5700, respectively. The localization error is calculated using Equation 4.1. Therefore, the 3D coordinates for true locations for both dipoles are first calculated. The true locations are 2000 and 5700, so the 3D coordinates corresponding to these locations are as follows:

$$\text{For } 2000 : [-59.22, -27.90, 25.71]$$
$$\text{For } 5700 : [45.99, -32.21, 25.79]$$
$$(9.2)$$

Furthermore, the results for the modified MSP with various numbers of patches are also produced; that is, after each step for optimization for hyperparameters, as explained in the previous chapter, the activation maps are produced. This error is compared with all previous algorithms results in the tables. Therefore, first, the results are plotted where patches are optimized at 300 and then at 700, and so on, until 1900 patches. The algorithm is run for multiple trials, and thus the data are averaged for free energy and the localization error for dipole at 2000 and dipole at 5700, respectively.

Hence, the generated source activity is measured using the algorithm for generation of synthetic data at the SNR level of 5 dB, and the resultant graph is shown in Figures 9.38 and 9.39.

In this way, the free energy and localization error for various algorithms are calculated for synthetic EEG data taken at one SNR level, 5 dB. However, the results from another SNR level are shown with their relative comparison for all techniques discussed above. These results are the average from multiple trials. Thus, the results from the first SNR level are summarized in Table 9.11. A detailed discussion is presented in the next section.

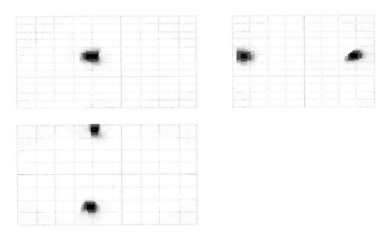

Figure 9.38 Generated source activity at signal-to-noise ratio (SNR) = 5 dB.

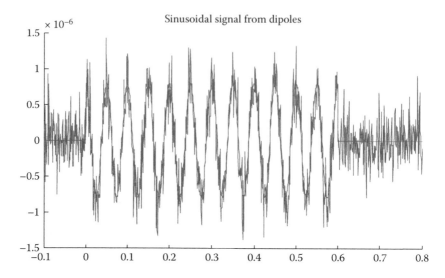

Figure 9.39 Sinusoidal signal from dipoles at signal-to-noise ratio (SNR) = 5 dB.

Table 9.11 Free Energy and Localization Error Comparison for Various Techniques

S. No.	Localization Method	Free Energy	Localization Error at Dipole Location 2000	Localization Error at Dipole Location 5700
01.	MSP	-839.12 ± 1.604	18.30 ± 20.50	64.23 ± 13.31
02.	LORETA	-933.03 ± 0.331	17.53 ± 6.7863	72.73 ± 1.474
03.	Minimum norm	-936.20 ± 0.715	20.24 ± 7.28	72.73 ± 6.48
04.	Beamforming	-910.60 ± 0.04	20.25 ± 0.50	72.73 ± 6.5
05.	Modified MSP (300 patches)	-807.04 ± 4.68	9.18 ± 0.297	9.05 ± 18.23
06.	Modified MSP (700 patches)	-796.29 ± 1.368	7.76 ± 0.90	7.76 ± 0.90
07.	Modified MSP (1100 patches)	-802.46 ± 2.092	7.45 ± 14.46	8.16 ± 2.487
08.	Modified MSP (1500 patches)	-802.57 ± 0.762	9.51 ± 3.98	16.53 ± 0.727
09.	Modified MSP (1900 patches)	-798.56 ± 2.02	7.76 ± 0.450	6.47 ± 0.352

LORETA, Low-resolution brain electromagnetic tomography; MNE, minimum norm estimation; MSP, multiple sparse priors.

9.5.3 Detailed discussion of the results from SNR = 5 dB

This section provides a detailed insight into the results obtained from the inversions of synthetic data generated at SNR = 5 dB. Some trends can be visualized from the data produced in Table 9.11 related to free energy, number of patches, and localization error for each inversion. Starting from the classical MNE and LORETA methods, the free energy values are −936.205 and −933.03, respectively. However, this value is very low as compared with MSP (−839.12) and different numbers of patches for the modified MSP algorithm. This low value shows less accuracy for the developed data. However, for the beamforming method, the free energy value is −910.6850, which is also low compared with MSP and modified MSP at different patches. Thus, Bayesian-based methods outperform in terms of free energy as compared with the classical minimum norm, LORETA, and beamforming approaches. However, the performance of the proposed modified MSP at different levels is better than MSP if compared in terms of free energy. It can be observed that the MSP has a free energy of −839.12, which is low as compared with all levels of modified MSP. The highest value of modified MSP was −796.29, where the number of patches is 700. Thus, the increased number of patches imparts improvement in accuracy, thereby increasing the free energy.

In terms of localization error for dipoles positioned at 2000 and 5700, the averaged error for MSP has error for dipole at 2000 (i.e., 18.3240) and also for dipole at 5700 (i.e., 64.23). Thus, when compared with LORETA (17.533 for dipole at 2000 and 72.73 for dipole at 5700), MNE (20.2465 for dipole at 2000 and 72.73 for dipole at 5700), and beamforming (30.023 for dipole at 2000 and 29.50 for dipole at 5700), respectively, it is revealed that MSP has slightly lower performance for the first dipole, that is, dipole at 2000, as compared with classical techniques except for beamformer where MSP has lower error for both dipoles. However, the error for dipole at 5700 for MSP is lower than all methods.

However, the localization error of modified MSP is far better as compared with MSP and classical algorithms. It can be seen that on average it is quite low. However, at some levels for dipole at 2000, it has very low values such as at 700 and 1100 patches, that is, 7.76 and 7.45, which is very low error as compared with the MSP and classical techniques. Continuing the discussion for the second dipole placed at the 5700 location, here also the modified MSP performs better than MSP and classical algorithms as it has the lowest localization error for all levels. This shows the superiority of modified MSP in terms of localization error over other existing algorithms.

9.5.4 Synthetic data results for SNR = 10 dB

In the second phase of synthetic data analysis, the SNR level is changed to 10 dB and all inversions are applied to analyze the change in the behavior.

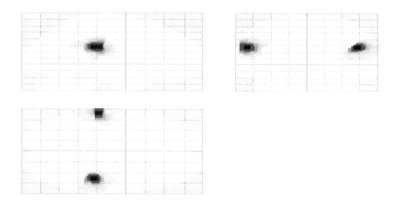

Figure 9.40 Generated source activity at signal-to-noise ratio (SNR) = 10 dB.

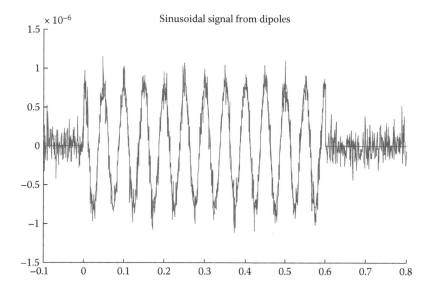

Figure 9.41 Sinusoidal signal from dipoles at signal-to-noise ratio (SNR) = 10 dB.

Hence, the results are presented for each algorithm in the form of averaged values for 40 trials conducted for each algorithm. A comparison table is produced to provide a comparison for various algorithms and the proposed modified MSP algorithm. The results for the modified MSP are produced for a change in the number of patches to show the change in localization error and free energy. The generated source activity is shown in Figure 9.40 and the generated dipole signal in Figure 9.41.

In the same way as done for SNR = 5 dB, the free energy and localization error for various algorithms are calculated for synthetic EEG data

Table 9.12 Free Energy and Localization Error Comparison for Various Techniques

S. No.	Localization Method	Free Energy	Localization Error at Dipole Location 2000	Localization Error at Dipole Location 5700
01.	MSP	−802.38 ± 1.169	19.98 ± 4.21	23.14 ± 9.94
02.	LORETA	−864.34 ± 4.088	17.79 ± 0.3128	37.16 ± 7.73
03.	Minimum norm	−865.28 ± 0.396	17.74 ± 0.412	37.16 ± 7.381
04.	Beamforming	−862.20 ± 1.09	30.02 ± 11.035	29.50 ± 7.986
05.	Modified MSP (300 patches)	−792.22 ± 2.649	15.26 ± 8.399	18.05 ± 0.944
06.	Modified MSP (700 patches)	−786.85 ± 2.33	8.75 ± 0.56	7.57 ± 0.256
07.	Modified MSP (1100 patches)	−791.17 ± 1.30	3.39 ± 7.94	16.20 ± 0.8796
08.	Modified MSP (1500 patches)	−788.55 ± 1.419	3.20 ± 1.60	15.0 ± 9.58
09.	Modified MSP (1900 patches)	−782.77 ± 0.543	7.76 ± 0.356	6.47 ± 0.312

LORETA, Low-resolution brain electromagnetic tomography; MSP, multiple sparse priors.

taken at this SNR level of 10 dB. However, results are shown in Table 9.12 with their relative comparison for all the techniques discussed above. These results are the average result from multiple trials. The results from the first SNR level are summarized in Table 9.12. A detailed discussion is presented in the following section.

9.5.5 Detailed discussion of the results from SNR = 10 dB

This section provides a discussion of the results obtained from the inversions of synthetic data generated at SNR = 10 dB. Some trends can be visualized from the data produced in Table 9.12 related to free energy, number of patches, localization error, and computational time for each inversion. Starting from the classical MNE and LORETA methods, where the free energy values are −864.34 and −865.28, respectively, this value is very low as compared with MSP and different number of patches for the modified MSP algorithm. This low value shows less accuracy for the developed data. However, for the beamforming method, the free energy value is −862.2, which is also low compared with the MSP and modified MSP at different patches. Thus, Bayesian-based methods outperform in terms of free energy as compared with the classical minimum norm, LORETA, and beamforming approaches. However, the performance of the modified MSP at different levels is better than the MSP if compared in terms of free energy. It can be seen that the MSP has free energy of 802.380, which is low as compared with all levels of modified MSP. The highest value free energy for the modified MSP was −782.775 where the number of patches is 1900. Thus, the increased number of patches imparts improvement in accuracy, thereby increasing the free energy. In terms of localization error for the dipoles positioned at 2000 and 5700, the MSP has an error for dipole at 2000 (i.e., 19.98) and also for dipole at 5700 (i.e., 23.140). Thus, when compared with LORETA (17.796 for dipole at 2000 and 37.162 for the dipole at 5700), MNE (17.744 for dipole at 2000 and 37.162 for dipole at 5700), and beamforming (30.023 for dipole at 2000 and 29.50 for dipole at 5700), respectively, it is revealed that MSP has slightly lower performance for the first dipole, at 2000, as compared with classical techniques except for beamformer where MSP has lower error for both dipoles. However, the error for dipole at 5700 for the MSP is lower than all methods. However, the localization error of the modified MSP is far better as compared with the MSP and classical algorithms. It can be seen that on average it is quite low. However, at some levels for dipole at 2000, it has very low values such as at 1100, 1500, and 1700 patches, that is, 3.395, 3.20, and 4.80, which are very low error as compared with MSP and classical techniques. Continuing the discussion for the second dipole placed at 5700 location, here also the modified MSP performs better than the MSP and classical algorithms as it has the lowest localization error for all levels.

9.5.6 Synthetic data results for SNR = 0 dB

In the next phase of synthetic data analysis, the SNR level is changed to 0 dB and all inversions are applied to analyze the change in the behavior. Hence, the results are presented for each algorithm in the form of averaged values for multiple trials conducted for each algorithm. A comparison table is produced to provide a comparison for various algorithms and the modified MSP algorithm. The results for the modified MSP are produced for a change in the number of patches to show the change in localization error and free energy. The generated source activity is shown in Figure 9.42 and the generated dipole signal in Figure 9.43.

In a similar fashion as described earlier, the free energy and localization error for various algorithms are calculated for synthetic EEG data taken at the new SNR level of 0 dB. However, results from this SNR level are shown in Table 9.13 with their relative comparison for all the aforementioned techniques. These results are average results from multiple trials. Thus, the results are summarized in Table 9.13. A detailed discussion is presented in the next section.

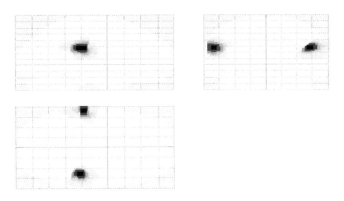

Figure 9.42 Generated source activity at signal-to-noise ratio (SNR) = 0 dB.

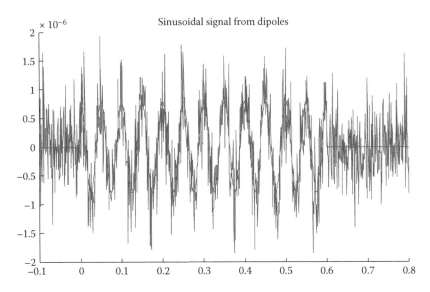

Figure 9.43 Generated dipole activity at signal-to-noise ratio (SNR) = 0 dB.

9.5.7 Detailed discussion of the results from SNR = 0 dB

This section provides a detailed insight into the results obtained from the inversions of synthetic data generated at SNR = 0 dB. The trends can be visualized from the data produced in Table 9.13 related to free energy, number of patches, localization error, and computational time for each inversion. Starting the discussion from the classical MNE and LORETA methods, where the free energy values are −1702.0 and −1726.30, respectively, this value is low as compared with MSP (−1551.20) and different numbers of patches for the modified MSP algorithm. The low values for classical techniques show

Table 9.13 Free Energy and Localization Error Comparison for Various Techniques

S. No.	Localization Method	Free Energy	Localization Error at Dipole Location 2000	Localization Error at Dipole Location 5700
01.	MSP	-1551.20 ± 24.084	28.92 ± 12.5	53.459 ± 8.96
02.	LORETA	-1726.30 ± 0.925	15.9774 ± 0.48	72.74 ± 0.301
03.	Minimum norm	-1702.0 ± 6.693	20.24 ± 0.58	73.45 ± 0.30
04.	Beamforming	-1692.10 ± 3.121	18.4971 ± 0.49	72.73 ± 0.39
05.	Modified MSP (300 patches)	-1485.90 ± 6.64	9.18 ± 0.35	14.33 ± 10.6
06.	Modified MSP (700 patches)	-1457.3 ± 2.275	7.76 ± 0.02	6.47 ± 0.15
07.	Modified MSP (1100 patches)	-1463.0 ± 3.516	5.6075 ± 0.18	10.548 ± 1.57
08.	Modified MSP (1500 patches)	-1471.2 ± 5.67	0.8626 ± 0.12	14.104 ± 0.45
09.	Modified MSP (1900 patches)	-1468.3 ± 4.48	7.76 ± 0.015	6.47 ± 0.15

LORETA, Low-resolution brain electromagnetic tomography; MSP, multiple sparse priors.

less accuracy for inverting the data. However, for the beamforming method, the free energy value is −1692.10, which is also low compared with MSP and modified MSP at different patches. Thus, Bayesian-based methods outperform in terms of free energy as compared with classical minimum norm, LORETA, and beamforming approaches. However, the performance of the modified MSP at different levels is better than the MSP if compared in terms of free energy. It is observed that the MSP has free energy of −1551.20, which is low as compared with all levels of modified MSP. The highest value of modified MSP was −1457.0, where the number of patches is 700. Hence, it is proved that even for lower values of SNR, modified MSP with increased number of patches performs better than MSP and classical techniques.

In terms of localization error, the MSP has average error for dipole at 2000 (28.92) and also for dipole at 5700 (53.459). Thus, when compared with LORETA (15.97 for dipole at 2000 and 72.74 for dipole at 5700), MNE (20.24 for dipole at 2000 and 73.45 for dipole at 5700), and beamforming (18.49 for dipole at 2000 and 72.73 for dipole at 5700), respectively, it is clear that MSP has slightly lower performance for the first dipole at 2000, as compared with classical techniques except for beamformer where MSP has lower error for both dipoles. However, the error for dipole at 5700 for MSP is lower than all methods.

Furthermore, the localization error of modified MSP is far improved as compared with MSP and classical algorithms. It can be seen that on average it is quite low. However, at some levels for dipole at 2000, it has very low values such as at 500, 1500, and 1700 patches, that is, 0.52, 0.8626, and 0.05, which is a very low error as compared with MSP and classical techniques.

9.5.8 Synthetic data results for SNR = −5 dB

Continuing the synthetic data analysis, the SNR level is changed to −5 dB and all inversions are applied to analyze the change in the behavior. Hence, the results are presented for each algorithm in the form of averaged values for multiple trials conducted for each algorithm. A comparison table is produced to provide a comparison for various algorithms and the proposed modified MSP algorithm. The results for modified MSP are produced for change in the number of patches to show the change in localization error and free energy. The generated source activity is shown in Figure 9.44 and the generated dipole signal in Figure 9.45.

In the same way as done for previous cases, the free energy and localization errors for various algorithms are calculated for synthetic EEG data taken at this SNR level of −5 dB. However, results from this analysis are shown in Table 9.13 with their relative comparison for all the aforementioned techniques. These are average results from multiple trials. A summary of the results is presented in Table 9.14, and a detailed discussion is given in the next section.

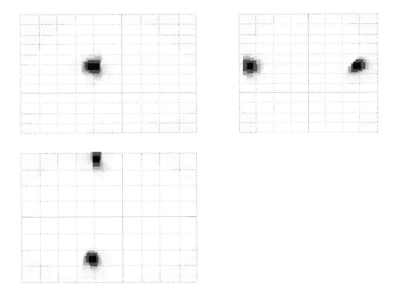

Figure 9.44 Generated source activity at signal-to-noise ratio (SNR) = −5 dB.

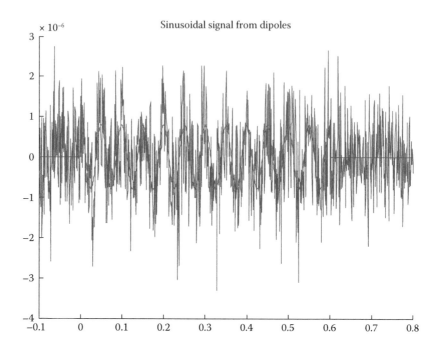

Figure 9.45 Sinusoidal signal from dipole at signal-to-noise ratio (SNR) = −5 dB.

Table 9.14 Free Energy and Localization Error Comparison for Various Techniques

S. No.	Localization Method	Free Energy	Localization Error at Dipole Location 2000	Localization Error at Dipole Location 5700
01.	MSP	−2833.6 ± 33.911	19.982 ± 1.58	72.51 ± 7.66
02.	LORETA	−3183.4 ± 4.11	15.98 ± 0.18	72.74 ± 9.88
03.	Minimum norm	−3036.3 ± 35.91	20.24 ± 0.18	73.75 ± 0.25
04.	Beamforming	−3092.8 ± 4.067	20.24 ± 0.12	73.75 ± 0.36
05.	Modified MSP (300 patches)	−2686.90 ± 5.779	18.802 ± 15.45	19.35 ± 18.61
06.	Modified MSP (700 patches)	−2651.3 ± 6.058	7.76 ± 0.205	6.47 ± 0.20
07.	Modified MSP (1100 patches)	−2651.70 ± 8.56	10.07 ± 3.49	8.49 ± 2.47
08.	Modified MSP (1500 patches)	−2641.8 ± 4.65	2.436 ± 2.10	13.850 ± 1.20
09.	Modified MSP (1900 patches)	−2668.8 ± 6.56	7.76 ± 0.20	6.47 ± 0.120

LORETA, Low-resolution brain electromagnetic tomography; MSP, multiple sparse priors.

9.5.9 Detailed discussion of the results from SNR = −5 dB

This section presents a discussion of the results obtained from the inversions of synthetic data generated at SNR = −5 dB. Starting the discussion from the classical MNE and LORETA methods, where the free energy values are −3036.3 and −3183.4, respectively, the value is low as compared with the MSP (−2833.6) and a different numbers of patches for the modified MSP algorithm. The low values for classical techniques show less accuracy for inversion. However, for the beamforming method, the free energy value is −3092.8, which is also low compared with MSP and modified MSP at different patches. Thus, Bayesian-based methods outperform in terms of free energy as compared with classical minimum norm, LORETA, and beamforming approaches. However, the performance of modified MSP at different levels is better than MSP if compared in terms of free energy. It can be seen that the MSP has free energy of −2833.6, which is low as compared with all levels of modified MSP. The highest value of modified MSP was −2641.8, where the number of patches is 1500. Hence, it is proved that even for lower values of SNR, modified MSP with increased number of patches performs better than MSP and classical techniques.

In terms of localization error, the MSP has an average error for dipole at 2000 (19.982) and also for dipole at 5700 (72.51). Thus, when compared with LORETA (15.98 for dipole at 2000 and 72.74 for dipole at 5700), MNE (20.24 for dipole at 2000 and 73.45 for dipole at 5700), and beamforming (20.24 for dipole at 2000 and 72.75 for dipole at 5700), respectively, it is clear that MSP has slightly lower performance for the first dipole at 2000 as compared with classical techniques except for beamformer, where MSP has lower error for both dipoles. However, the error for a dipole at 5700 for MSP is lower than all methods. Furthermore, the localization error of the modified MSP is far better as compared with MSP and classical algorithms. It can be seen that on average it is quite low.

9.5.10 Synthetic data results for SNR = −20 dB

In the final phase of synthetic data analysis, the SNR level is changed to −20 dB with different amplitude level as mentioned in the beginning of the chapter. Thus, all inversion methods are applied to see the change in the behavior. Hence, the results are presented for each algorithm in the form of averaged values for multiple trials conducted for each algorithm. A comparison table is produced to provide a comparison for various algorithms and the modified MSP algorithm. The results for the modified MSP are produced for a change in the number of patches to show the change in localization error and free energy. The generated source activity is shown in Figure 9.46 and the generated dipole signal in Figure 9.47.

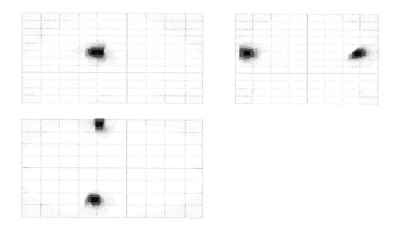

Figure 9.46 Generated source activity at signal-to-noise ratio (SNR) = −20 dB.

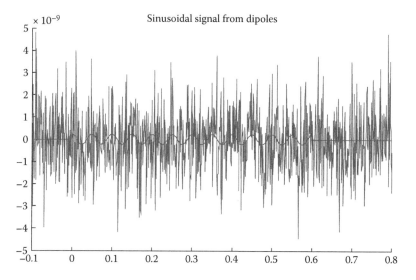

Figure 9.47 Sinusoidal signal from dipole at signal-to-noise ratio (SNR) = −20 dB.

In a similar way as described before, the free energy and localization error for various algorithms are calculated for synthetic EEG data taken at a new SNR level, that is, −20 dB. Hence, the results from this SNR level are shown in Table 9.13 with their relative comparison for all techniques mentioned above. These are average results from multiple trials, and the results are summarized in Table 9.15. A detailed discussion is presented in the next section.

This section discusses the results obtained from the inversions of synthetic data generated at SNR = −20 dB. Starting the discussion from

Table 9.15 Free Energy and Localization Error Comparison for
Various Techniques

S. No.	Localization Method	Free Energy	Localization Error at Dipole Location 2000	Localization Error at Dipole Location 5700
01.	MSP	−10,530 ± 20.30	9.18 ± 0.25	30.03 ± 21.88
02.	LORETA	−10,793 ± 29.91	14.25 ± 1.25	41.098 ± 3.264
03.	Minimum norm	−10,794 ± 32.76	15.25 ± 1.25	41.098 ± 3.264
04.	Beamforming	−10,793 ± 31.85	9.20 ± 0.5185	39.079 ± 2.56
05.	Modified MSP (300 patches)	−10,528 ± 16.68	8.194 ± 1.264	16.43 ± 2.26
06.	Modified MSP (700 patches)	−10,499 ± 36.58	7.76 ± 0.30	6.47 ± 0.25
07.	Modified MSP (1100 patches)	−10,517 ± 32.31	8.176 ± 0.25	9.12 ± 0.30
08	Modified MSP (1500 patches)	−10,514 ± 39.30	1.6 ± 2.06	15.34 ± 2.79
09.	Modified MSP (1900 patches)	−10,496 ± 35.60	4.76 ± 0.45	6.47 ± 0.30

LORETA, Low-resolution brain electromagnetic tomography; MSP, multiple sparse priors.

the classical MNE and LORETA methods, where the free energy values are −10,794 and −10,793, respectively, this value is low as compared with MSP (−10,530) and a different number of patches for the modified MSP algorithm. The low values for classical techniques show less accuracy for inversion. However, for the beamforming method, the free energy value is −10,793, which is also low compared with the MSP and modified MSP at different patches. However, the performance of the proposed modified MSP at different levels is better than the MSP if compared in terms of free energy. It can be seen that the MSP has a free energy of −10,530, which is low as compared with all levels of the modified MSP. The highest value of the modified MSP observed was −10,496 where the number of patches is 1900. Hence, it is proved that even for lower values of SNR and different dipole amplitudes, the modified MSP with an increased number of patches performs better than MSP and classical techniques.

In terms of localization error, the MSP has an average error for dipole at 2000—that is, 9.18 and also for a dipole at 5700 as 30.03. Thus, when compared with LORETA (14.25 for dipole at 2000 and 41.098 for dipole at 5700), MNE (15.25 for dipole at 2000 and 41.098 for dipole at 5700), and beamforming (9.20 for dipole at 2000 and 39.079 for dipole at 5700), respectively, it is clear that MSP has better performance for both dipoles. In addition, the localization error of modified MSP is far better as compared with MSP and classical algorithms. It can be seen that on average it is quite low.

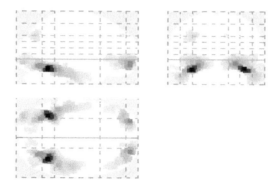

Figure 9.48 Best case for the modified multiple sparse priors (MSP) (5 dB).

From this discussion and observations made for various inversion techniques, it is evident that the localization capability of an inversion technique is improved in terms of low localization error and high free energy when the patches are increased. Hence, the algorithms were run for multiple trials to assert the fact statistically. Therefore, in this section just the best cases of modified MSP are produced for each SNR level (Figures 9.48 through 9.52).

In addition, some of the brain maps from multiple runs of MSP, LORETA, MNE, and beamformer are produced in Figures 9.53 through 9.56.

It can be seen from the comparison of best cases for the modified MSP and classical techniques that modified MSP has got more accuracy in terms of localization. The inversion result for the modified MSP is much closer to that of synthetic generated dipole activity as shown in figures for various noise levels. Thus, the classical techniques have blurring effects and not clear source visibility. Hence, the modified MSP performs the best for various SNR levels.

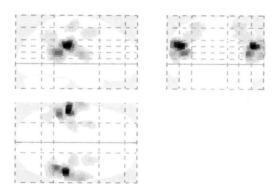

Figure 9.49 Best case for the modified multiple sparse priors (MSP) (10 dB).

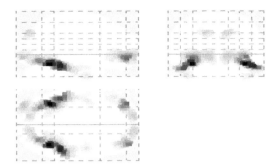

Figure 9.50 Best case for the modified multiple sparse priors (MSP) (0 dB).

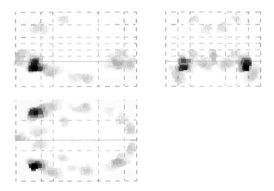

Figure 9.51 Best case for the modified multiple sparse priors (MSP) (−5 dB).

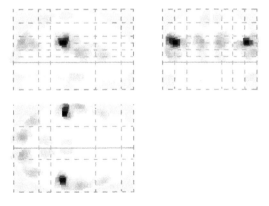

Figure 9.52 Best case for the modified multiple sparse priors (MSP) (−20 dB).

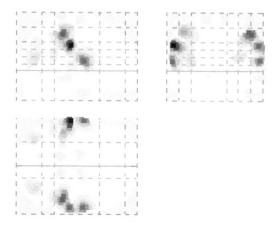

Figure 9.53 Multiple sparse priors (MSP) result from random trial.

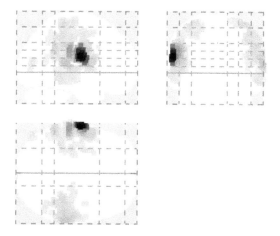

Figure 9.54 Low-resolution brain electromagnetic tomography (LORETA) activation map from a random trial.

9.6 Reduced channel source localization

The concept of using less channels (electrodes) for the source estimation is not novel. This concept is elaborated in many research articles for suggesting less channels to yield better solutions. In Michel et al. [2], it is explained that the proper sampling of the spatial frequencies of the electric fields associated with brain activity can lead to a better resolution of the topographic features. In the literature, it is also suggested that the interelectrode distance should be around 2–3 cm for avoidance of any distortion related to scalp electric potential distributions [3,4].

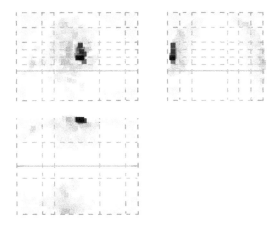

Figure 9.55 Minimum norm estimation (MNE) activation map from a random trial.

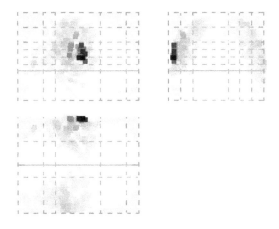

Figure 9.56 Beamformer activation map from a random trial.

Thus, some previous studies have estimated the sources with classical LORETA, LAURA, and MNE methods for real-time and synthetic EEG data with varying numbers of electrodes. However, there are some methods that have suggested recursively applied and projected-multiple signal classification (RAP-MUSIC) in conjunction with a developed reduced conductivity method for source localization [5,6]. In this research work, a unique approach is adopted to help the researchers and neuroscientists use the least number of electrodes with optimized accuracy expressed as free energy and less computational complexity. Thus, the developed approach is tested using real-time EEG data and synthetic EEG data with all classical algorithms (LORETA, MNE, and

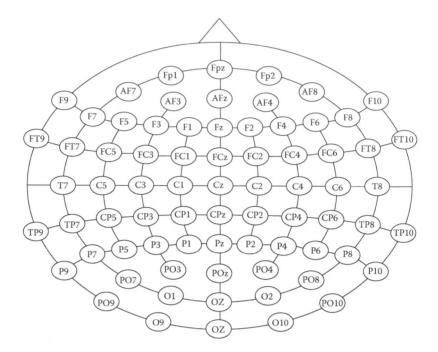

Figure 9.57 Easycap EEG electrode layout.

beamformer) and Bayesian framework-based MSP and proposed modified MSP, respectively. The process is outlined as follows:

- In the first step, the electrodes that are used to acquire the EEG
  data are mapped into the Enobio EEG electrode layout [7]. The EEG
  was recorded using a 74-channel Easycap EEG cap [8]. The standard
  10%–10% electrode system was applied for readings. The electrode
  layout for the Easycap system is shown in Figure 9.57. This electrode
  cap is mapped into the Enobio cap, which has only eight electrodes.
  The Enobio layout is shown in Figure 9.58. The green channels are
  those that are selected for analysis only.
- The next step is selecting the green electrodes in the layout shown
  in Figure 9.58 and making all the remaining electrodes as "other" in
  MATLAB-based SPM environment. This will reduce the number of
  electrodes from 74 to 7 as the effect of "*Pz*" is considered merged into
  "*Cz*" as they are located in the same zone with no electrode between
  them.
- The SPM object is created with less electrodes and is thus loaded into
  for inversion. Hence, at the first instance, MSP is applied followed by
  application of classical algorithms, which includes LORETA, MNE,

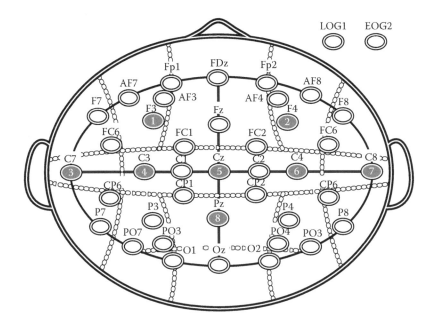

Figure 9.58 ENOBIO electrode layout.

and beamformer. The free energy and computational time for each algorithm are saved to compare with the modified MSP algorithm.
- Finally, the modified MSP is applied for this new dataset. Thus, 10 instances are applied to analyze the behavior of modified MSP in terms of free energy and computational time.
- The results from various algorithms are summarized in the form of figures and tables.

The Easycap electrode layout and its mapping into the Enobio layout are shown in Figures 9.57 and 9.58, respectively.

9.6.1 *Results for MNE, LORETA, beamformer, MSP, and modified MSP for synthetic data*

The procedure for reduced electrode localization is repeated for synthetic EEG data. Thus, two electrodes are used for the generation of EEG data with positions at 2000 and 5700, respectively. The number of electrodes is 7 as was done before for real-time EEG data. The SNR level is kept at 5 dB. The generated source activity is shown in Figures 9.59 and 9.60.

Thus, the inversions are applied for multiple trials and the results are produced in the form of a table. However, the results are averaged for

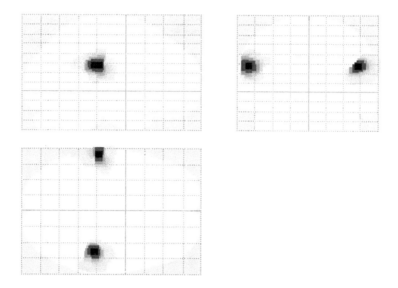

Figure 9.59 Generated source activity for seven-electrode synthetic data.

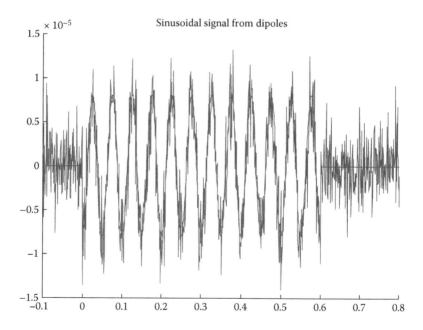

Figure 9.60 Sinusoidal signal generated from synthetic data at signal-to-noise ratio (SNR) = 5 dB.

multiple trials, and so some of the random inversion maps are presented for comparison between various techniques (Figures 9.61 through 9.65).

The localization error is calculated using the same formula as that used for synthetic data before (Table 9.16).

From the comparison table, it can be seen that again modified MSP has performed best as compared with classical algorithms and Bayesian-based MSP, respectively. The average difference between the free energy as provided by the modified MSP and classical algorithms is 10 units,

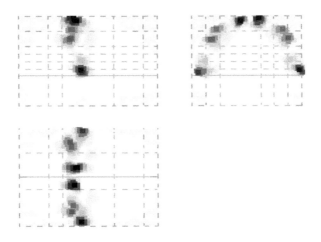

Figure 9.61 Multiple sparse priors (MSP) results from a random trial.

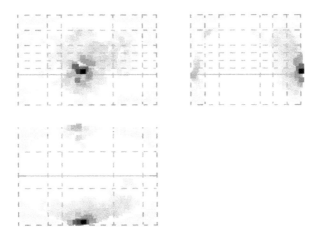

Figure 9.62 Low-resolution brain electromagnetic tomography (LORETA) activation map for a random trial.

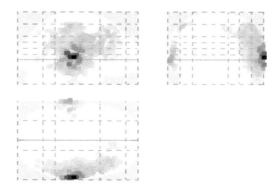

Figure 9.63 Minimum norm estimation (MNE) activation map for a random trial.

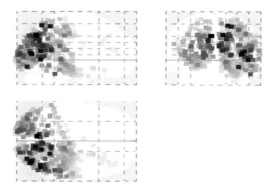

Figure 9.64 Beamformer activation map for a random trial.

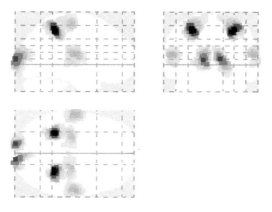

Figure 9.65 Modified multiple sparse priors (MSP) activation map for a random trial (*Best Case: 1100 Patches*).

Table 9.16 Free Energy and Localization Error Comparison for
Various Techniques

S. No.	Localization Method	Free Energy	Localization Error at Dipole Location 2000	Localization Error at Dipole Location 5700
01.	MSP	-88 ± 20.25	36.71 ± 1.58	33.58 ± 6.54
02.	LORETA	-94.6 ± 13.08	69.29 ± 3.56	60.80 ± 7.34
03.	Minimum norm	-93.82 ± 15.0	68.59 ± 4.25	61.20 ± 5.93
04.	Beamforming	-94.5 ± 31.02	66.25 ± 7.85	61.20 ± 8.10
05.	Modified MSP (300 patches)	-84.7 ± 1.25	33.0 ± 1.56	28.86 ± 2.025
06.	Modified MSP (700 patches)	-84.4 ± 2.056	32.41 ± 1.025	31.95 ± 2.256
07.	Modified MSP (1100 patches)	-79.9 ± 0.785	12.85 ± 1.025	15.67 ± 0.64
08.	Modified MSP (1500 patches)	-81.0 ± 2.35	61.57 ± 0.25	67.45 ± 0.54
09.	Modified MSP (1900 patches)	-84.8 ± 0.625	35.28 ± 0.35	32.75 ± 0.564

LORETA, Low-resolution brain electromagnetic tomography; MSP, multiple sparse priors.

which is very large when we are dealing with such a small number of electrodes—that is, seven as generated using the reduced number of electrodes. Thus, the modified MSP performs best for this novel technique with a reduced number of channels.

9.6.2 Real-time EEG data and reduced channel results

After producing the results with reduced channels for synthetic EEG data generated at the SNR level of 5 dB, now the results are generated for real-time EEG data for reduced sensor levels. Thus, the classical methods and modified MSP are applied to see the behavior of the change in free energy with each change in the number of patches. However, Table 9.17 shows the free energy and computationals time for classical, MSP, and modified MSP methods observed with reduced numbers of electrodes.

From the comparison table, it can be seen that again modified MSP has performed the best as compared with classical algorithms and Bayesian-based MSP, respectively. The average difference between the free energy as provided by modified MSP and classical algorithms is 150 units, which is very large when we are dealing with such a small number of electrodes—that is, seven as generated using the reduced number of channels. Thus, the free energy difference for MSP and modified MSP is on average 20 units, which is a significant difference for small electrode data.

Table 9.17 Real-Time EEG Data Comparison for Seven Electrodes

S. No.	Inversion Method	Free Energy	Computational Time (s)
01.	MSP	−771.3	5.3245
02.	LORETA	−893.4	4.3678
03.	MNE	−893.4	4.2658
04.	Beamformer	−861.8	10.3652
05.	Modified MSP (300 patches)	−769.0	5.3394
06.	Modified MSP (500 patches)	−755.3	6.1414
07.	Modified MSP (700 patches)	−768.8	7.3673
08.	Modified MSP (900 patches)	−756.7	8.5830
09.	Modified MSP (1100 patches)	−760.9	9.9799

LORETA, Low-resolution brain electromagnetic tomography; MNE, minimum norm estimation; MSP, multiple sparse priors.

Summary

This chapter provides a detailed discussion on various aspects of the results obtained for EEG data inversion through various classical and new techniques. The results are divided into two main categories: from synthetic data or from real-time EEG data, respectively. The synthetic data were observed for different values of SNR levels. Thus, the analysis is carried out for all existing methods and the proposed modified MSP. A detailed discussion is provided for all methods where these methods are compared in terms of free energy, localization error (only for synthetic data), and computational time. A similar methodology was followed for real-time EEG data where the number of subjects was kept at 10. It was concluded from real-time and synthetic data (generated at different SNR levels) that modified MSP perform far better with respect to free energy, localization error, and computational complexity.

The localization was observed for reduced electrodes with simple mapping of 74 electrodes into seven electrodes only. However, with the reduced number of electrodes, the free energy is optimized as seen in the results. Again, the modified MSP is compared with classical and MSP algorithms in terms of free energy and computational complexity. Thus, it was revealed that modified MSP outperformed in both of these parameters as compared with classical algorithms and MSP, respectively.

References

1. Statistical Parametric Mapping. Available: http://www.fil.ion.ucl.ac.uk/spm/. Accessed on May 1, 2017.

2. C. M. Michel, M. M. Murray, G. Lantz, S. Gonzalez, L. Spinelli, and R. G. de Peralta, EEG source imaging, *Clinical Neurophysiology*, vol. 115, pp. 2195–2222, 2004.

3. S. Ahlfors et al., Spatiotemporal activity of a cortical network for processing visual motion revealed by MEG and fMRI, *Journal of Neurophysiology*, vol. 82, pp. 2545–2555, 1999.

4. G. Anogianakis et al., A consensus statement on relative merits of EEG and MEG, *Electroencephalography and Clinical Neurophysiology*, vol. 82, pp. 317–319, 1992.

5. B. R. Yitembe, G. Crevecoeur, R. Van Keer, and L. Dupre, Reduced conductivity dependence method for increase of dipole localization accuracy in the EEG inverse problem, *IEEE Transactions on Biomedical Engineering*, vol. 58, pp. 1430–1440, 2011.

6. B. R. Yitembe, G. Crevecoeur, R. Van Keer, and L. Dupre, EEG inverse problem solution using a selection procedure on a high number of electrodes with minimal influence of conductivity, *IEEE Transactions on Magnetics*, vol. 47, pp. 874–877, 2011.

7. ENOBIO. Available: http://www.neuroelectrics.com/products/enobio/. Accessed on January 18, 2017.

8. EASYCAP. Available: http://www.easycap.de/. Accessed on January 18, 2017.

chapter ten

Future directions for EEG source localization

Introduction

This chapter summarizes the main theme of this book as well as potential contributions from this research work. Hence, the future work is included to give directions for researchers in this area for better results and application in healthcare problems. This work is carried out for brain source estimation using electroencephalography (EEG) as the neuroimaging technique. In this conjunction, the inversion methods are used to localize the active sources due to which the electromagnetic activity is produced inside the brain. The inversion methods are reviewed thoroughly in terms of the parameters provided by them; thus, they were implemented on real-time EEG data and synthetically generated EEG data, respectively. The behavior of the methods is compared in terms of a cost function, which is termed as free energy, computational time, and the localization error, respectively. The higher values of free energy are needed to prove the efficiency of the algorithm used, as free energy defines the accuracy with respect to changes brought in an inversion algorithm. Thus, localization error is another parameter that is desired to be as low as possible. A reduced electrode model is proposed and implemented for various inversions to show the validity of the developed algorithm for a varying number of channels for real-time EEG data. Thus, a comparison is provided in terms of free energy and computational burden for all methods and the proposed method.

It is evident from the results and observations made that with the increase in the number of patches within the Bayesian framework, the localization efficiency is improved. This positive change was demonstrated through improved free energy and minimized localization error, respectively. However, there exists room for further improvement in the techniques in terms of computational time and localization error. These improvements are summarized in the following section.

10.1 Future directions

This section describes the future directions for this research work. However, it should be noted that these future directions are provided on the basis of the observations made during this research work. These are open research questions, so the research community in this research field is invited to discuss further. Some of the research directions are as follows:

1. *Varying the distribution assumption*
   The probability distribution is assumed to be Gaussian throughout this research work. However, the distribution can be altered, which can bring about some basic changes in cost function—that is, free energy. This might be a contribution for research in source estimation as a new cost function might optimize the results.
2. *Validation of reduced electrodes technique with various datasets*
   The technique with reduced electrodes is used for real-time EEG data, which shows the performance in terms of free energy and computational complexity. However, taking advantage of the basic theory provided in this work, the method can be further implemented for synthetic data at various signal-to-noise ratio (SNR) levels and with various noise levels to check the behavior of algorithms (including modified multiple sparse priors) at various levels. Thus, the results can be extended to varying channel levels, for example, 12, 25, 50, to see the behavior of the trends between various parameters and, thus, produce a novelty in the field of channel reduction for source estimation.
3. *Results with various head modeling schemes*
   It has been reported in various publications as well as discussed in previous chapters that proper head modeling yields a solution with better resolution and less localization error. However, this is possible by comparing results obtained through various head modeling schemes, which include finite difference method, finite element method, and boundary element method. Hence, it is suggested to confirm the results for real as well as synthetic data with multiple runs to validate the results statistically.
   Besides these recommendations, there are many changes that can be introduced to improve the localization ability. Some are addressed here, which include the inclusion of anatomical features into analysis, forward modeling constraints, varying conductivity modeling, prior component definition framework, and so on.

10.2 Significance of research with potential applications

The research work for the field of brain source localization using EEG signals has significant applications for various medical and clinical applications. The work is carried out under various conditions—that is, low SNR and high SNR levels with various amplitude levels. Thus, the testing is done for free energy, localization error, and computational time. The experiment for a reduced number of electrodes was also carried out to show the significance of the developed method—that is, modified multiple sparse priors for low-cost applications. Thus, the method can be used for designing a source localization system for a number of healthcare applications. Some potential applications are as follows:

- The designed algorithm can be used by hospitals to help surgeons and physicians operating on patients with brain disorders.
- This algorithm can be used by researchers in the field of neuroscience in particular and signal processing experts in general.
- A product can be commercialized from the designed algorithm, which can be launched in the neuromarket so that hospitals, researchers, and experts can get optimum benefit from it.

Appendix A: List of software used for brain source localization

1. Netstation, EGI, US. Available: https://www.egi.com/research-division/geodesic-eeg-system-components/eeg-software. Accessed on August 5, 2017.
2. EEGLAB, Swartz Center for Computational Neuroscience, University of California, US. Available: https://sccn.ucsd.edu/eeglab/. Accessed on August 5, 2017.
3. Statistical Parametric Mapping, University College London, UK and Wellcome Trust Centre for Neuroimaging, UK. Available: http://www.fil.ion.ucl.ac.uk/spm/. Accessed on August 5, 2017.
4. s/e/LORETA, University of Zurich, Switzerland. Available: http://www.uzh.ch/keyinst/loreta.htm. Accessed on August 5, 2017.
5. Brainstorm, University of Southern California, US. Available: http://neuroimage.usc.edu/neuro/BrainStorm. Accessed on August 5, 2017.
6. Fieldtrip, Donders Institute of Brain, Cognition and Behavior, The Netherlands. Available: http://www.fieldtriptoolbox.org/. Accessed on August 5, 2017.
7. Brain Electrical Source Analysis (BESA), Germany. Available: http://www.besa.de/. Accessed on August 5, 2017.
8. Electromagnetic Brain Activity Localisation Software (EMBAL), INRIA, France. Available: http://embal.gforge.inria.fr/. Accessed on August 5, 2017.
9. OpenMEEG, INRIA Sophia-Antipolis, France. Available: https://openmeeg.github.io/# Accessed on August 5, 2017.
10. Neuroelectromagnetic Forward Modeling toolbox (NFT), Swartz Center for Computational Neuroscience, University of California, US. Available: https://sccn.ucsd.edu/wiki/NFT. Accessed on August 5, 2017.
11. SCIRun, University of Utah, US. Available: https://www.sci.utah.edu/cibc-software/scirun/forward-inverse-toolkit.html. Accessed on August 5, 2017.

12. Neuronic Source Analyzer, CNeuro, Cuba. Available: http://www.neuronicsa.com/. Accessed on August 5, 2017.

13. ERPLAB. Available: http://www.erpinfo.org/. Accessed on August 5, 2017.

14. SimBio, Kyoto University, Japan. Available: http://www.sim-bio.org/. Accessed on August 5, 2017.

15. BioMesh3D, University of Utah, US. Available: https://www.sci.utah.edu/cibc-software/scirun/biomesh3d.html. Accessed on August 5, 2017.

Appendix B: Pseudocodes for classical and modern techniques

Pseudocode for modified multiple sparse priors (MSP) implementation protocol

Input: Preprocessed EEG data $D \in \Re^{380 \times 991 \times 3}$, gain/lead-field matrix

 Output: Inversion of EEG data at certain co-ordinates in 3D activation map

 Define:

Time window of interest ←[start time stop time]

Smoothening factor (SF) for source priors ←$(0 \le SF \le 1)$

Number of patches ← 100

Low pass and high pass frequency ← any value that suits

Standard deviation for Gaussian temporal correlation ← 4

Checking lead field and optimization for spatial projector

   for i=1 to length(D)

      check the lead field and get number of dipoles and channels

      check for null spaces and remove them

   endfor

Average lead-field computation

   for $i = 1$ to number of modes

      initialize average lead field and compute regularized inverses for lead field

      for $j = 1$ to 8

         eliminate redundant virtual channels

         eliminate low signal-to-noise ratio (SNR) spatial modes by $SVD(UL \times UL^T) \ni UL =$ reduced lead field

         normalize the lead field as $UL = UL/$scaling factor

      endfor

   endfor

Temporal projector evaluation
for $i = 1$ to length(D)
   define time window of interest
   apply Hanning operator and filtering
endfor
Optimize spatial priors over subjects
  Creating spatial basis in source space
  for $i = 1$ to *number of patches*
     calculate modified source space for left hemisphere ∋
$S_{MSP(LH)} = K_{mod} \times q$ and q is derived from smoothening Green's
function
     calculate for the right hemisphere depending on *vert*
matrix
     calculate for bilateral
endfor
Inversion
if OP-MSP with greedy search is selected **then**
  number of patches ← length of source space
    for $i = 1$ to *number of patches*
        define new source space by modification introduced
earlier
      endfor
Calculate multivariate Bayes, which provides Bayesian
optimization of multivariate linear model
Empirical priors are accumulated
  elseif OP-MSP with automatic relevance determination (ARD)
is selected **then**
      Use restricted maximum likelihood (ReML) to
estimate the covariance components from sample covariance
matrix
Design spatial priors using the number of patches and
modified hyperparameters
for $i = 1$ to *number of patches*
    if hyperparameter lies within threshold **then**
      define new source space based on modified
hyperparameters and spatial priors
   endif
endfor
    accumulate empirical priors based on modified lead-
field UL and spatial prior
    Optimized patches
for   number of patches 200 to 2100
         all the parameters are updated
         observe the results
           if Optimization of hyperparameters is
successful in terms of free energy and localization error **then**
           Plot the solution and compare with
previous solution

```
            else  Optimize the patches and repeat from
previous step
                    endif
        endfor
Plotting
    Plot the solution and observe the free energy and
localization error
if     free energy is high
        localization error is low endif
else start again optimizing the parameter
```

Pseudocode for implementation protocol of MSP (ARD and greedy search)

```
Input: Preprocessed EEG data D ∈ ℜ³⁸⁰ˣ⁹⁹¹ˣ³, gain/lead-field
matrix
Output: Inversion of EEG data at certain co-ordinates in 3D
activation map
  Define:
Time window of interest ← [start time stop time]
SF for source priors ←(0 ≤ SF ≤ 1)
Number of patches ← 100
Low pass and high pass frequency ← any value that suits
Standard deviation for Gaussian temporal correlation ← 4
Checking lead field and optimization for spatial projector
        for i = 1 to length(D)
            check the lead field and get number of dipoles and
channels
            check for null spaces and remove it
endfor
Average lead-field computation
    for i = 1 to number of modes
            initialize average lead field and compute
regularized inverses for lead field
                for j = 1 to 8
                    eliminate redundant virtual channels
                    eliminate low SNR spatial modes by
SVD(UL×ULᵀ) ∋ UL = reduced lead field
                        normalize the lead field as UL = UL/scaling
factor
                endfor
endfor
Temporal projector evaluation
for i = 1 to length(D)
    define time window of interest
    apply Hanning operator and filtering
endfor
```

Optimize spatial priors over subjects
  Creating spatial basis in source space
  for $i = 1$ to *number of patches*
    calculate modified source space for left hemisphere ∋
$S_{MSP(LH)} = K_{mod} \times q$ and q is derived from smoothening Green's
function
    calculate for the right hemisphere depending on *vert*
matrix
    calculate for bilateral
  endfor
Inversion
if MSP with greedy search is selected **then**
  number of patches ← length of source space
    for $i = 1$ to *number of patches*
        define new source space by modification introduced
earlier
    endfor
Calculate multivariate Bayes, which provides Bayesian
optimization of multivariate linear model
Empirical priors are accumulated
  elseif MSP with ARD is selected **then**
        Use ReML to estimate the covariance components from
sample covariance matrix
Design spatial priors using number of patches and modified
hyperparameters
for $i = 1$ to number of patches
    if hyperparameter lies within threshold **then**
        define new source space based on modified
hyperparameters and spatial priors
    endif
endfor
accumulate empirical priors based on modified lead-field UL
and spatial prior
Plotting
    Plot the solution and observe the free energy and
localization error.
if free energy is high
    localization error is low **endif**
else start again optimizing the projectors and repeat
through Step 2

Pseudocode for implementation protocol for minimum norm estimation (MNE)

Input: Preprocessed EEG data $D \in \Re^{380 \times 991 \times 3}$, gain/lead-field
matrix

Output: Inversion of EEG data at certain co-ordinates in 3D
activation map
  Define:
Time window of interest ←[start time stop time]
SF for source priors ←$(0 \leq SF \leq 1)$
Number of patches ← 100
Low pass and high pass frequency ← any value that suits
Standard deviation for Gaussian temporal correlation ← 4
*Checking lead field and optimization for spatial
projector*
        for $i = 1$ to length(D)
            check the lead field and get number of dipoles and
channels
            check for null spaces and remove it
    endfor
Average lead-field computation
        for $i = 1$ to number of modes
            initialize average lead field and compute
regularized inverses for lead field
                for $j = 1$ to 8
                    eliminate redundant virtual channels
                    eliminate low SNR spatial modes by
SVD$(UL \times UL^T)$ ∋ $UL =$ reduced lead field
                normalize the lead field as $UL = UL$/scaling
factor
        endfor
endfor
Temporal projector evaluation
for $i = 1$ to length(D)
    define time window of interest
    apply Hanning operator and filtering
endfor
Optimize spatial priors over subjects
Create source component
create minimum norm prior based on modified lead-field UL
and source component.
Inversion
for $i = 1$ to number of patches
            Define spatial priors as done in previous steps
            Accumulate empirical priors
endfor
  Plotting
      Plot the solution and observe the free energy and
localization error.
if   free energy is high
        localization error is low **endif**
else start again optimizing the projectors and repeat
through Step 2

Pseudocode for implementation protocol for low-resolution brain electromagnetic tomography (LORETA)

Input: Preprocessed EEG data $D \in \Re^{380 \times 991 \times 3}$, gain/lead-field matrix
Output: Inversion of EEG data at certain co-ordinates in 3D activation map
Define:
Time window of interest ←[start time stop time]
SF for source priors ←$(0 \leq SF \leq 1)$
Number of patches ← 100
Low pass and high pass frequency ← any value that suits
Standard deviation for Gaussian temporal correlation ← 4
Checking lead field and optimization for spatial projector
      for $i = 1$ to length(D)
          check the lead field and get number of dipoles and channels
          check for null spaces and remove it
      endfor
Average lead-field computation
  for $i = 1$ to number of modes
      initialize average lead field and compute regularized inverses for lead field
        for $j = 1$ to 8
          eliminate redundant virtual channels
          eliminate low SNR spatial modes by
$SVD(UL \times UL^T) \ni UL =$ reduced lead field
          normalize the lead field as $UL = UL/$scaling factor
        endfor
endfor
Temporal projector evaluation
for $i = 1$ to length(D)
   define time window of interest
   apply Hanning operator and filtering
endfor
Optimize spatial priors over subjects
create minimum norm prior based on modified lead-field UL and source component
add smoothness component in source space
$\ni S_{LOR} = K_{\mathrm{mod}} \times Qp \times K_{\mathrm{mod}}^{T}$

Inversion
for $i = 1$ to number of patches
        Define spatial priors as done in previous steps
        Accumulate empirical priors
endfor

```
 Plotting
     Plot the solution and observe the free energy and
localization error.
if free energy is high
   localization error is low endif
else start again optimizing the projectors and repeat
through Step 2
```

Pseudocode for implementation protocol for beamforming

Input: Preprocessed EEG data $D \in \Re^{380 \times 991 \times 3}$, gain/lead-field
matrix
Output: Inversion of EEG data at certain co-ordinates in 3D
activation map
Define:
Time window of interest ←[start time stop time]
SF for source priors ← $(0 \leq SF \leq 1)$
Number of patches ← 100
Low pass and high pass frequency ← any value that suits
Standard deviation for Gaussian temporal correlation ← 4
Checking lead field and optimization for spatial projector

```
     for i = 1 to length(D)
          check the lead field and get number of dipoles and
channels
          check for null spaces and remove it
     endfor
```
Average lead-field computation
```
  for i = 1 to number of modes
           initialize average lead field and compute
regularized inverses for lead field
          for j = 1 to 8
               eliminate redundant virtual channels
               eliminate low SNR spatial modes by
SVD(UL×UL^T) ∋ UL = reduced lead field
               normalize the lead field as UL = UL/scaling
factor
          endfor
endfor
```
Temporal projector evaluation
```
for i = 1 to length(D)
   define time window of interest
   apply Hanning operator and filtering
endfor
```
Optimize spatial priors over subjects
```
for i = 1 to number of sources
   normalized power is calculated using modified lead-field UL
   source power is calculated
```

     power for all sources is calculated by dividing source
power to normalized power
endfor
    The power from all sources is normalized to generate
source component $\ni S_{Beamformer} = K_{mod} \times Qp \times K_{mod}^{T}$
Inversion
for $i = 1$ to number of patches
         Define spatial priors as done in previous steps
         Accumulate empirical priors
endfor
  Plotting
     Plot the solution and observe the free energy and
localization error.
if free energy is high
  localization error is low **endif**
else start again optimizing the projectors and repeat
through Step 2

Index

T - #0419 - 071024 - C3 - 234/156/11 - PB - 9780367884970 - Gloss Lamination